Seismic Isolation for Architects

Seismic isolation offers the highest degree of earthquake protection to buildings and their inhabitants. Modern applications of the technology are less than 50 years old and uptake in seismically active regions continues to soar.

Seismic Isolation for Architects is a comprehensive introduction to the theory and practice in this field. Based on the latest research findings and the authors' extensive experience, coverage includes the application, effectiveness, benefits, and limitations of seismic isolation, as well as the architectural form, design aspects, retrofitting, economics, construction, and maintenance related to this method.

The book is written for an international audience: the authors review codes and practices from a number of countries and draw on examples from 11 territories including the USA, Chile, Argentina, Italy, Japan, and New Zealand. Aimed at readers without prior knowledge of structural engineering, the book provides an accessible, non-technical approach without using equations or calculations, instead using over 200 drawings, diagrams, and images to support the text. This book is key reading for students on architecture and civil engineering courses looking for a clear introduction to seismic isolation, as well as architects and engineers working in seismically active regions.

Andrew Charleson is an Associate Professor at the School of Architecture, Victoria University of Wellington, New Zealand. A structural engineer with many years' earthquake engineering experience, he now teaches Structures. He has previously authored two books, *Seismic Design for Architects* and *Structure as Architecture.*

Adriana Guisasola is an architect and Associate Professor in Structures at the Faculty of Architecture, Urban Planning and Design, Mendoza University, Argentina. She is in the process of completing her doctoral thesis on 'Architecture and Base Isolation'. She has authored many papers for world conferences on earthquake engineering and seismic isolation, energy dissipation, and active vibration control of structures.

Between them, the authors have been actively involved in the design and construction of three seismically isolated buildings and have visited and studied over 60 more in 11 different countries.

"I believe that this very well written and documented book will be very useful to architects worldwide. In fact, it explains the features and advantages of seismic isolation in a very useful way to architects, by stressing the fact that, thanks to this technique, it is possible not only to make buildings much safer at limited additional construction costs (if any), but also to allow for adopting architectural solutions that could never be applicable to conventionally founded buildings." – *Alessandro Martelli, President of the Italian Association GLIS, Founding President and present Vice-President of the Anti-Seismic Systems International Society (ASSISi), and former Professor of "Constructions in Seismic Areas" at the Faculty of Architecture of the University of Ferrara, Italy*

Seismic Isolation for Architects

Andrew Charleson and Adriana Guisasola

Routledge
Taylor & Francis Group
LONDON AND NEW YORK

First published 2017
by Routledge
2 Park Square, Milton Park, Abingdon, Oxon OX14 4RN

and by Routledge
711 Third Avenue, New York, NY 10017

Routledge is an imprint of the Taylor & Francis Group, an informa business

British Library Cataloguing-in-Publication Data
A catalogue record for this book is available from the British Library

Library of Congress Cataloging-in-Publication Data
Names: Charleson, Andrew, author. | Guisasola, Adriana, author.
Title: Seismic isolation for architects / Andrew Charleson and Adriana Guisasola.
Description: Abingdon, Oxon; New York: Routledge, 2017. |
Includes bibliographical references and index.
Identifiers: LCCN 2016008766| ISBN 9781138813205 (hardback: alk. paper) |
ISBN 9781138813212 (pbk.: alk. paper) | ISBN 9781315748313 (ebook)
Subjects: LCSH: Earthquake resistant design. | Isolating mechanisms. |
Buildings – Vibration.
Classification: LCC TA658.44 .C37 2017 | DDC 693.8/52 – dc23
LC record available at http://lccn.loc.gov/2016008766

ISBN: 978-1-138-81320-5 (hbk)
ISBN: 978-1-138-81321-2 (pbk)
ISBN: 978-1-315-74831-3 (ebk)

Typeset in Minion Pro and Avenir
by Florence Production Ltd, Stoodleigh, Devon, UK

– CONTENTS –

– ACKNOWLEDGEMENTS –

WE thank all those who have helped us along the journey towards the publishing of this book. In particular, the photographic and computer technicians at Victoria University of Wellington Schools of Architecture and Design are thanked for their support. Also thanks to Duncan McNaughton for drawing all the diagrams, and to the individuals and companies who have provided images. Unacknowledged images are provided by the authors.

– 1 –

Introduction

Purpose and timing of the book

MANY books and articles on the seismic isolation of buildings have been written for structural engineers. However, these publications are highly technical in nature and therefore are unsuitable for the vast majority of those who design, construct, own, insure and inhabit buildings in seismically active regions. The purpose of this book, then, is to introduce a relatively new game-changing technology to a wide audience concerned about how buildings and their contents are affected by earthquakes.

A research and writing project such as this, which seeks to address all issues relevant to seismic isolation, could have been undertaken at any time over the last 35 years. Yet it would have inevitably left unanswered many important questions relevant to the seismic isolation of buildings. In particular, how confident can we be in the ability of seismic isolation to reduce damage during earthquakes, and second, how does this relatively new approach to protect buildings from earthquakes compare to more conventional ones? Is seismic isolation really worth adopting?

Within a period of 18 days ending on 11 March 2011, the answers to these questions suddenly became much clearer. On 22 February the city of Christchurch experienced a devastating earthquake located only 10 km from its centre, and at a shallow depth of 5 km. While only one base-isolated building in Christchurch, the Christchurch Women's hospital, was tested by the earthquake, so was the entire building stock of Christchurch. Hundreds of buildings, many designed in accordance with one of the world's most advanced seismic codes, survived without collapse. But tragically, for different reasons, most have subsequently been demolished. This situation raises considerable uncertainty regarding the appropriateness of modern philosophies of seismic design.

Then, on the 11 March, Japan was struck by the massive Magnitude 9 Tōhoku Earthquake, centred off the east coast. Most of the destruction was caused by the tsunami that destroyed coastal areas, but earthquake shaking damaged many buildings, and for the first time, tested hundreds of seismically isolated buildings on an unprecedented scale. Although subjected to a lesser intensity of shaking than they had been designed for, these buildings performed very well. The two earthquakes occurred within days but were located thousands of kilometres apart along the Pacific tectonic plate. Together they both demonstrated both the effectiveness of seismic isolation as well as deficiencies in current design approaches to earthquake attack, accentuating the benefits of seismic isolation.

Current approaches to the seismic design of buildings

BEFORE examining seismic isolation in detail, a primary aim of this book, we need to first explain the vulnerability of most buildings around us to earthquakes, and then explore the strengths and weaknesses of conventional as well as new structural concepts to protect buildings against earthquakes.

The buildings we inhabit tend to instil a false sense of safety, particularly with respect to earthquakes. A number of factors converge to soothe our fears about the seismic safety of a building. We note how buildings stand without any visible signs of distress. They support their self-weight plus the loads of all the equipment, stock and other contents their occupants impose. It is rare for buildings to collapse when gravity is the only force acting. We also correctly understand that the chances of an earthquake occurring during a given period of time, perhaps today or even during our lifetime, range from infinitesimal to minor. We might also believe that because the buildings around us have experienced one or more earthquakes during their lives, they can therefore be assumed safe. People who are more knowledgeable about building construction might believe that because buildings are built using modern materials of reinforced concrete or structural steel, they are safe.

Most of these perceptions are only partially correct. The reality is that the risk of a building collapsing during an earthquake is far higher than one might imagine. The unfortunate fact is that most buildings in the world, unless they have been seismically retrofitted or built since the early 1980s, are unsafe in earthquakes. In some countries even more modern buildings are at-risk. The vulnerability of unreinforced masonry buildings is well known. The undesirable combination of heavy and brittle material and a lack of strong tying between walls, and floors to walls means these buildings are usually the first to suffer earthquake damage. But what is less well known is the hazard more modern buildings pose.

It was only after the mid-1970s that codes of practice internationally rectified a critical flaw in the way buildings were designed against earthquakes. It was then that the importance of *ductility* was recognized – that ability of the structure of a building to survive earthquake over-load without collapse yet suffering damage. Although buildings designed prior to this period did possess adequate strength, for what would by today's standards be understood as small earthquakes, in many cases they will partially or fully collapse when the shaking of a large earthquake exceeds their strength (Figures 1.1 and 1.2). Rather than their structural elements such as beams, columns or structural walls bending, stretching or shearing in a *ductile* manner during a large earthquake, they break or snap suddenly, in a *brittle* way.

1.1 Example of structural damage in a building with inadequate shear walls at the garage level. Loma Prieta earthquake, San Francisco USA, 1989

Source: US Geological Survey

1.2 The seriously damaged Olive View Hospital moment frame building. San Fernando Valley earthquake, California USA, 1971

Source: National Oceanic & Atmospheric Administration (NOAA)

If such buildings survive a damaging earthquake without at least partial collapse, it will not be because of how the primary structural members were designed and detailed. Surviving buildings will have likely benefited from, for example, additional strength provided by walls required for fire protection along site boundaries or to confined stairwells, as well as structurally desirable characteristics such as symmetry and regularity.

Today, modern codes require that buildings avoid collapse when their primary seismic-resisting structural systems experience overload during an earthquake. And overload *is* likely given how low design load levels are where compared to loads occurring during design earthquakes. Typical modern earthquake-resistant building design begins with the selection of one of three common structural systems (Figures 1.3 and 1.4). One system will resist horizontal earthquake forces in one direction in the plan of the building, and the same or another structural system will act in the direction at 90 degrees to it. The numbers and sizes of structural members depend on many factors including the size, height and weight of the building, and the level of seismicity of the site. With strength in *both* orthogonal directions, the structure can cope with earthquake attack from *any* direction. Where parallel structural systems, such as shear walls, are separated in plan, so they don't act along the same line, they prevent the building from excessive twisting.[1] Then the structural engineer calculates the sizes and details of the structural members so that they are either so strong as not to be overloaded in an earthquake or, more commonly, designs them to be ductile. That is, even if a member is damaged it maintains most of its strength and doesn't break. But, as graphically illustrated by the Christchurch earthquake, ductility means damage which can lead to demolition.

Even before the limitations of current earthquake-resistant design approaches were so starkly revealed, engineers had been developing new devices such as dampers, rather like car shock absorbers, and modifying existing structural systems to survive medium to large earthquakes with little or no structural damage. The use of dampers and 'damage-free' systems are growing in popularity, but are still uncommon (Figures 1.5 and 1.6). These new systems certainly represent a big improvement over conventional systems in which structural damage is inevitable during a large earthquake, but they are unable to offer enhanced protection for architectural elements such as claddings, interior fit outs and building contents. Seismic isolation alone protects both structure and architectural elements.

Introducing seismic isolation

SEISMIC isolation is a term describing any sort of isolation, including by far its most common form, base isolation. Most buildings are isolated at their bases, but occasionally the isolation plane is located in a middle storey. Seismic isolation involves the partial separation of a building from the ground underneath it to reduce the intensity of the earthquake shaking it experiences. Seismic isolation is like placing the superstructure of a building on the surface of foundations that have been oiled or greased to make them slippery.

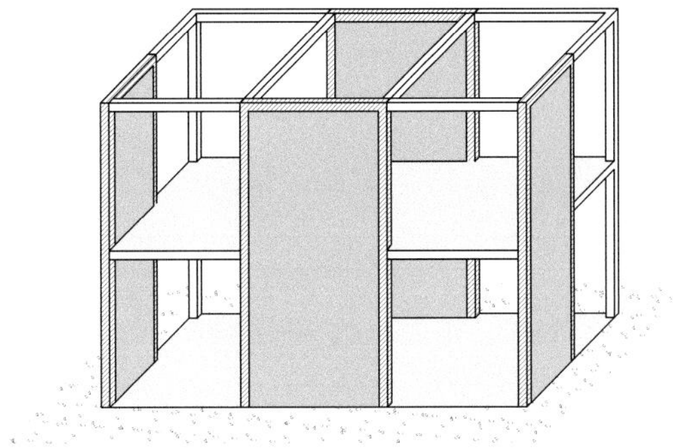

(a) Shear walls

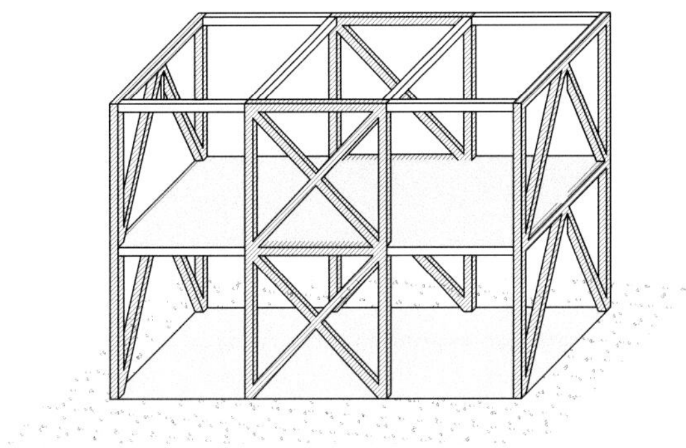

(b) Braced frames

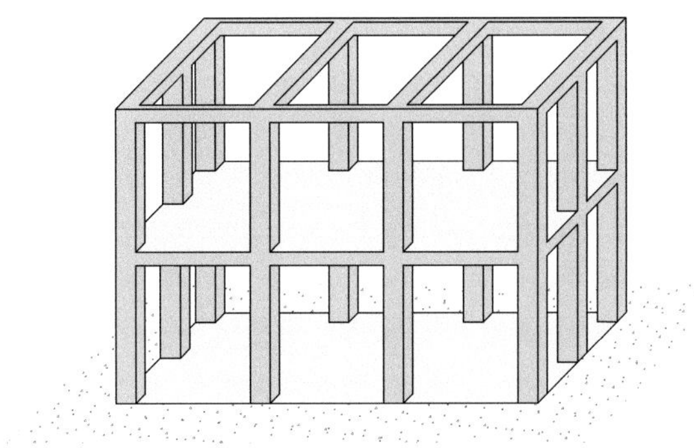

(c) Moment frames

1.3 The three most common structural systems for resisting horizontal forces

1.4a Example of a type of structural system: shear wall

When the most damaging earthquake waves shake the ground horizontally back and forth, the superstructure is protected by the sliding interface. If only it were that simple in practice! Rather than grease, elastomeric and or sliding bearings form the slippery isolation plane. The dynamic impact of a building superstructure against perimeter retaining walls is avoided by constructing physical separation gaps (Figure 1.7).

Seismically isolating a building is rather like moving a ship from a dry dock, where it is resting on the ground, and launching it into water. Once released from its foundations and floating on elastomeric or other bearings, an isolated structure is far less vulnerable to the effects of horizontal ground shaking. Usually, most of the energy of an earthquake is manifest in high frequency vibration. But since the bearings of a seismically isolated building are very flexible for horizontal movement, the building doesn't resonate like a conventional fixed-base building. Rather than experiencing violent high frequency shaking at its base, which is then amplified up its height, an isolated building enjoys a far gentler ride and an absence of increased accelerations and movement on its upper floors (Figure 1.8). Its low frequency response means less earthquake energy is transferred into the superstructure. This reduces or even eliminates damage to structural members, architectural elements and building contents as exemplified in the following two well-documented case studies.

1.4b Example of a type of structural system: single-bay moment frame

1.4c Example of a type of structural system: eccentrically braced frame

1.5 A damping device on the bottom left incorporated into cross-bracing in the National Museum of Emerging Science and Innovation, Tokyo, Japan

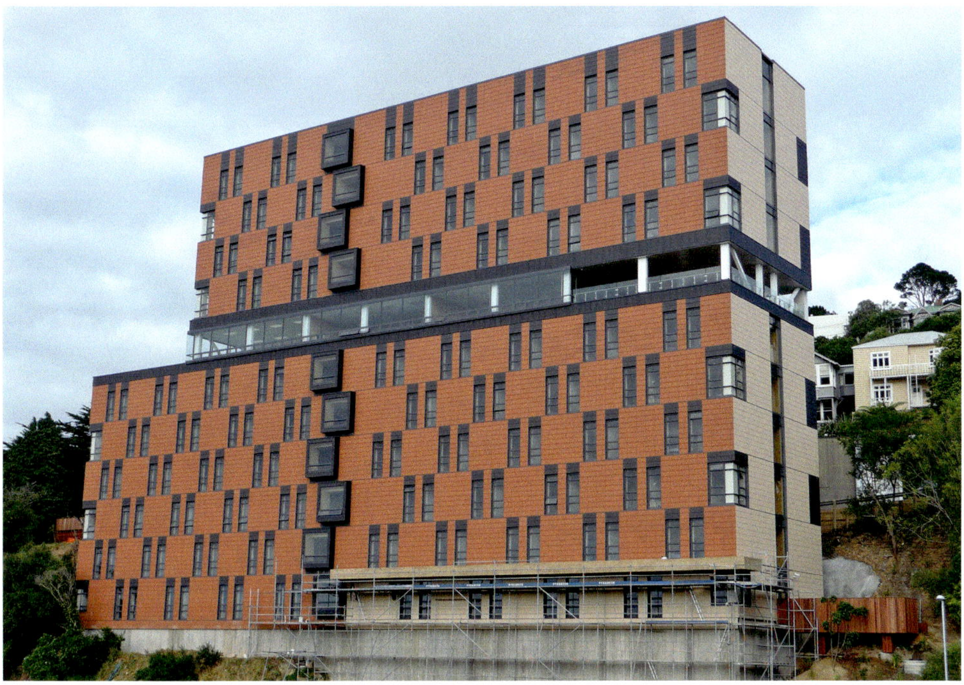

1.6 A damage-free structure in a hostel at Victoria University of Wellington, New Zealand. In the event of seismic overload, structural steel connections in the moment frames (along the building) and cross-braced frames (across), slip rather than break

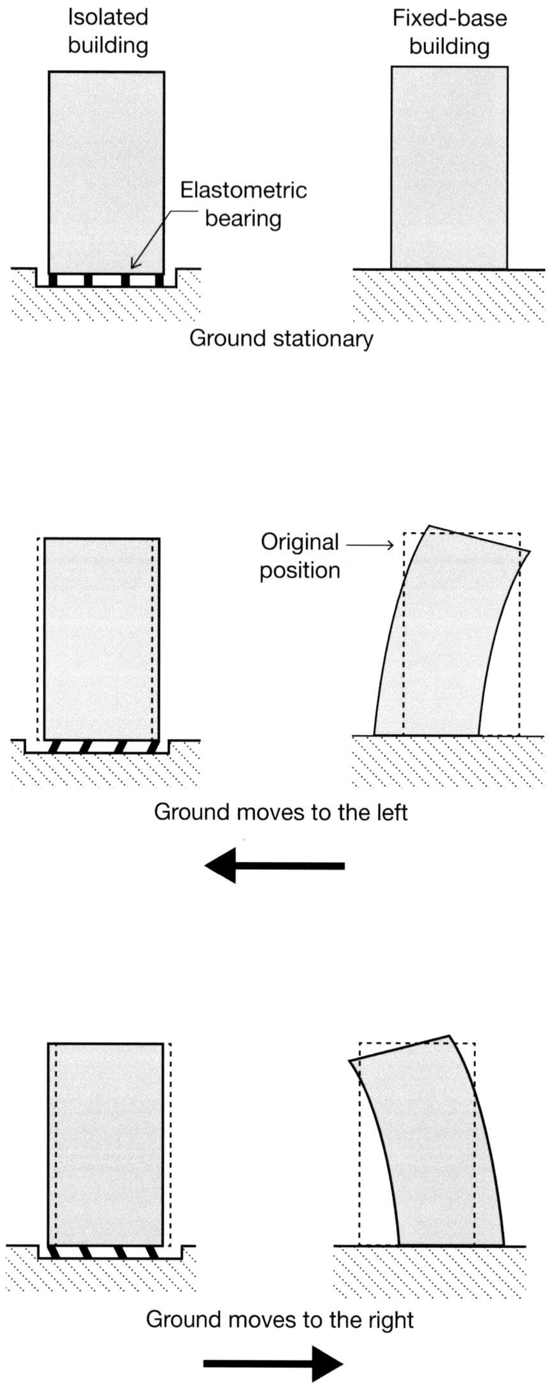

1.7 The essence of seismic isolation. An isolated building is protected from earthquake shaking by moving at its base. Movement in a conventional building is absorbed up its height

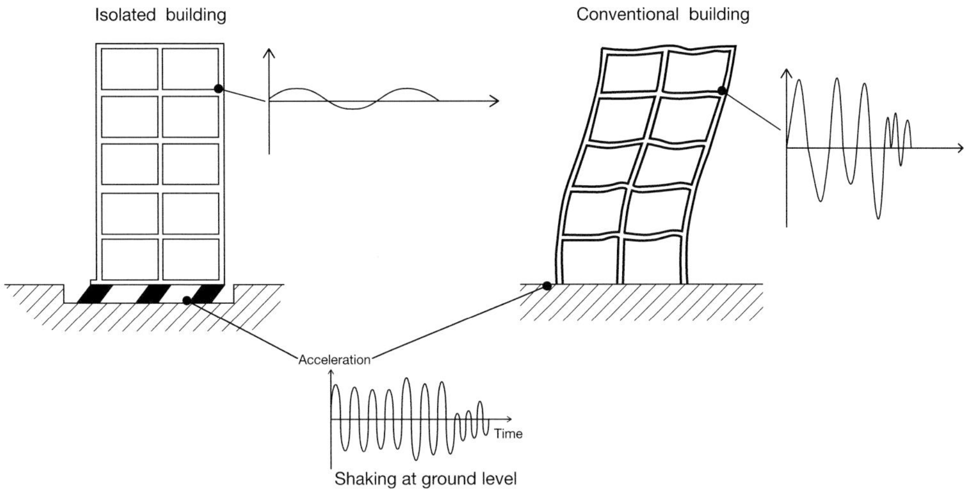

1.8 The different responses of seismically isolated and fixed-base buildings

Tale of two hospitals

We begin by recounting the response of the Ishinomaki Red Cross Hospital to the 2011 Tohoku earthquake. The 6-storey 420 bed hospital was built in 2006 (Figure 1.9). It is located 4.5 km from the east coast of Japan, just far enough inland to escape the tsunami. It was the closest hospital to the earthquake epicentre, some 100 km away. Fortunately, the effect of the earthquake on the hospital was filmed. The video begins with scenes in an office during the building shaking and then shows the immediate preparation for the arrival of the injured.[2] In spite of its seismic isolation, the intensity of shaking within the building was considerable. Some workers remained at their desks while others steadied vibrating computer monitors and mobile equipment. Apart from papers and files falling from desks and shelves, no damage was observed to the structure or architectural finishes. Thanks to this excellent seismic performance and the reliable back-up electricity generation, hospital functioning was unaffected. Nearby, buildings designed to the most recent 1981 Japanese code suffered slight structural damage, but major damage to building contents, suspended ceilings and external cladding.[3] During the quake the hospital moved to-and-fro relative to the ground beneath it by up to 260 mm, approximately half the design earthquake movement.[4]

Christchurch Women's Hospital was opened in 2005 (Figure 1.10). Consisting of one main basement level and seven storeys above ground, earthquake forces in the superstructure are resisted by reinforced concrete moment frames in the longitudinal direction (Figure 1.9). Steel V-braced frames, in the lowest four levels, combine with full-height concrete moment frames to resist transverse loads. The hospital rests mainly on lead-rubber bearings, designed to undergo plus or minus 420 mm horizontal displacement during the design earthquake expected, on average, every 2000 years.

1.9 Ishinomaki Red Cross Hospital, Miyagi, Japan

Source: Fuji-s

1.10 Christchurch Women's Hospital, Christchurch, New Zealand

Source: Nabil Allaf

On 4 September 2010 it experienced a moderate earthquake which did little damage to surrounding buildings, although many unreinforced masonry buildings were badly damaged. During this quake the bearings displaced only 40 mm but staff reported trolleys rolling along the floor, items falling from shelves and the need to hang on to stop falling during the 'furious side-to-side swaying'.[5] No damage other than to sacrificial components covering seismic gaps was reported.

That earthquake proved to be minor compared to the one of 22 February 2011. This time most buildings near the hospital were badly damaged. Some 50 per cent of the reinforced concrete frame buildings in the CBD were posted with yellow placards, allowing only brief periods of occupancy, and 21 per cent were so badly damaged that occupancy was prohibited. Of buildings of the same era as the Christchurch Women's Hospital, approximately 50 per cent suffered moderate to serious structural damage, and 135 people died when two mid-rise concrete buildings collapsed. Over 1000 commercial buildings have subsequently been demolished.[6] The intensity of ground shaking was approximately twice that assumed in the New Zealand seismic design code, although far shorter in duration.[7]

During this catastrophic event the lead-rubber bearings of the Christchurch Women's Hospital displaced approximately 200 mm, protecting the building from significant damage. Minor damage included cracking in some partitions and cladding around window openings, its basement being flooded with liquefied soil from outside its retaining walls, services pipes damaging a fire wall and, finally, roof-top 'chillers moved around and piping for the condenser collapsed'.[8] The absence of any structural damage and the minimal damage to architectural elements and services set it apart from the other buildings in the hospital precinct. An oncology unit from a damaged building was relocated into it shortly after the earthquake.

Reality check

ALTHOUGH these two case studies illustrate outstanding seismic performance of seismically isolated structures as compared to those with conventional fixed-bases, there is no sense in which seismic isolation is a miracle solution. As noted above, particularly in smaller earthquakes, the shaking inside a building can be vigorous. Although only a small proportion of the earthquake energy at the base of the building is transferred into its superstructure, structure such as walls, moment or cross-braced frames are still required to resist horizontal inertia forces within the superstructure. Unfortunately for architects and building owners, the structural members of these systems are of similar dimensions to those in conventional buildings. This is because seismic design codes allow the design forces of conventional ductile buildings to be reduced by factors between four and six in acknowledgement of the ductility they are designed for, to prevent collapse. Therefore, the structures of seismically isolated buildings are designed to similar levels of force, but the really important difference is that they remain undamaged during the design earthquake. Conventional structures, on the other hand, will be seriously damaged and, as sadly observed in Christchurch, may be demolished.

This remarkably improved seismic performance usually comes at a price. A cost premium for quality is inherent in every purchase we make. For example, a study of three base-isolated hospitals constructed since 2008 showed that the extra cost of the isolation system was approximately three per cent of the total construction cost.[9] Although the economics of seismic isolation is discussed in Chapter 10, we note in passing that the investment in the isolation system of the Christchurch Women's Hospital has already been recouped many times over. It prevented costly damage and ensured operational continuity immediately after the earthquake.

The following chapters of this book examine in far more detail the topics raised above and many others beside. We attempt to discuss every aspect of seismic isolation of relevance to the reader by anticipating your questions and answering them.

Notes

1 Charleson, A.W., 2008, *Seismic design for architects: outwitting the quake*, Oxford: Elsevier.

2 Japanese Red Cross Society 2011, 'Ishinomaki Red Cross Hospital – recording of the Great East Japan Earthquake', video recording. Available from: www.youtube.com/watch?v=Pc1ZO7YwcWc (accessed: 30 December 2015).

3 Motosaka, M. & Mitsuji, K., 2012, 'Building damage during the 2011 off the Pacific Coast of Tohoku earthquake', *Soils and foundations*, vol. 52, no. 5, pp. 929–44.

4 Someya, T., 2013, 'Seismically isolated hospital offers ray of hope in disaster – Ishinomaki Red Cross Hospital', Proceedings of the 13th World Conference on Seismic Isolation, Energy Dissipation and Active Vibration Control of Structures, 24–27 September, Sendai, Japan, p. 3.

5 Gavin, H.P. & Wilkinson, G., 2010, 'Preliminary observations of the effects of the 2010 Darfield earthquake on the base-isolated Christchurch Women's hospital', *Bulletin of the New Zealand Society for Earthquake Engineering*, vol. 43, no. 4, 360–7, p. 362.

6 Whittaker, D., 2013, 'Recent developments in seismic isolation in New Zealand', Proceedings of the 13th World Conference on Seismic Isolation, Energy Dissipation and Active Vibration Control of Structures, 24–7 September, Sendai, Japan, p. 1.

7 Kam, W.Y., Pampanin, S. & Elwood, K., 2010, 'Seismic performance of reinforced concrete buildings in the 22 February Christchurch (Lyttelton) earthquake', *Bulletin of the New Zealand Society for Earthquake Engineering*, vol. 44, no. 4, pp. 239–77.

8 McIntosh, J.K., Jacques, C., Mitrani-Reiser, J. *et al.*, 2012, 'The impact of the 22nd February 2011 earthquake on Christchurch hospital', Proceedings of the 2012 Technical Conference of the New Zealand Society for Earthquake Engineering, p. 6.

9 Charleson, A.W. & Allaf, N.J., 2012, 'Costs of base-isolation and earthquake insurance in New Zealand', Proceedings of the 2012 Technical Conference of the New Zealand Society for Earthquake Engineering, Christchurch, p. 6.

– 2 –

History, requirements and principles of seismic isolation

Brief history

ALTHOUGH seismic isolation as practiced today has a short history of less than 50 years, evidence of far earlier attempts to isolate buildings from ground shaking are being discovered. Detailing for isolation is found within the vernacular architecture of some earthquake-prone communities. Some far older traditional buildings have been observed to outperform modern buildings. For example, some ancient Iranian buildings are founded on smooth flat stones or on sand layers under load-bearing walls, presumably to facilitate sliding. A most compelling example from that region consists of houses supported on short lengths of logs which are layered and oriented in plan to allow rolling in two orthogonal directions.[1] These buildings performed well during the 1990 Manjil earthquake. They moved up to 200 mm on their foundations without damage.

The first modern application of seismic isolation was in an elementary school in Skopje, Yugoslavia, founded on neoprene (artificial rubber) bearings in 1979. However the William Clayton Building, Wellington, completed in 1981, was the first building whose isolation system included all the five necessary characteristics of seismic isolation, as explained later in this chapter (Figure 2.1). Seismic isolation was achieved using lead-rubber bearings, newly invented by Dr Bill Robinson.

After those initial applications, the uptake internationally of seismic isolation was slow.[2] Then the 1995 Kobe, Japan, earthquake struck, causing hundreds of buildings to collapse and damaging thousands more. Only 10 seismically isolated buildings per year had been constructed before that earthquake; after the earthquake, the rate surged to 150 buildings per year between 1995 and 2002.[3] Renewed interest in and accelerated up-takes of seismic isolation have been observed in other countries as well more recently. This has been due to a combination of loss of life during an earthquake, a huge amount of building damage, and the excellent observed seismic performance of one or more local seismically isolated buildings. This accelerated application of seismic isolation was also evident in China after the 2008 Wenchuan earthquake, in Italy after the 2009 Abruzzo earthquake, and in New Zealand after the 2011 Christchurch earthquake.

2.1 The William Clayton Building, New Zealnd, Wellington, 1981. The first example of the comprehensive application of seismic isolation

A recent international overview of seismically isolated buildings reports that as at 2012 over 6600 buildings in Japan were seismically isolated.[4] This number includes 4000 houses. China had over 2500 buildings of which 70 per cent are houses;[5] the Russian Federation, 550 buildings (and bridges); Italy, 300; and the USA, approximately 200. There are also up to several tens of seismically isolated buildings in other countries including South Korea, Taiwan, Armenia, Chile and New Zealand. The numbers continue to increase world-wide. Several examples of seismically isolated buildings from numerous countries are shown in Figures 2.2 to 2.20.

The types of seismically isolated buildings in terms of their function are extremely diverse. Before exploring some of their typologies it is worth noting the reasons why owners isolate their buildings. In his lectures on seismic isolation, Ron Mayes summarizes them as follows:

- 'emergency response,
- continued business operations,
- protect contents,
- reduce damage repair cost,
- protect architecture, and
- occupant safety – "peace of mind".'

It therefore comes as no surprise to encounter seismically isolated hospitals or emergency response centres. In any community these should be the most highly seismically protected

2.2 Military Hospital La Reina, Santiago, Chile, 2007

Source: J.C. De La Llera

2.3 Angels of San Giuliano School, San Giuliano di Puglia, Italy, 2008

2.4 Student residence, National Technological University, Mendoza Regional Faculty, Mendoza, Argentina, 2007

Source: M. Tornello

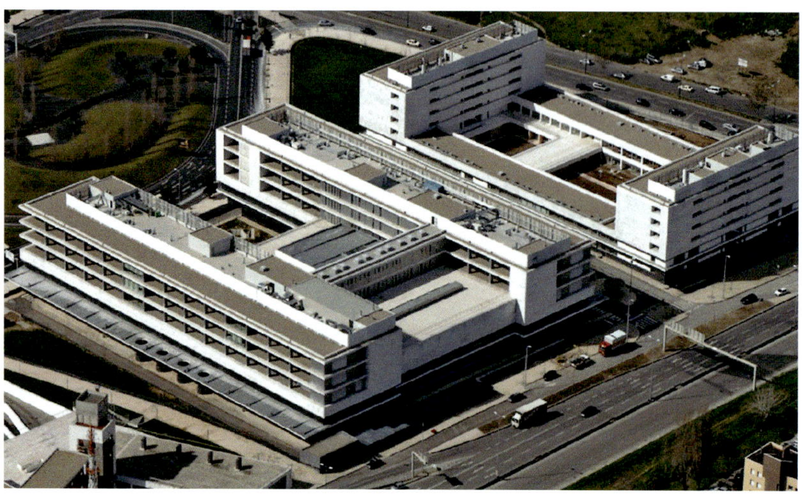

2.5 Hospital of Light, Lisbon, Portugal

Source: FIP Industriale S.p.A., Italy

2.6 Institute of Histology and Embryology of Mendoza (IHEM) laboratories, Mendoza, Argentina, 2014

Source: Contractor Santiago Monteverdi CC SA

2.7 Acropolis Museum, Athens, Greece, 2009

2.8 Part of one of 185 residential buildings constructed after the 2009 L'Aquila earthquake, Italy, as part of the CASE project. Multi-storey housing was built upon seismically isolated ground floor slabs above basements

Source: FIP Industriale S.p.A., Italy

2.9 The car parking basement of a CASE project building. Double concave curved surface sliders isolate the ground floor slab from the steel cantilever columns

Source: FIP Industriale S.p.A., Italy

2.10 Glendale City Hall, Glendale, USA. The interior courtyard showing one of the isolators used to retrofit the original 1927 building

Source: USGS

2.11 San Francisco City Hall, San Francisco, USA, retrofitted 1999

buildings of all. Then there are buildings housing manufacturing or other functions for whom downtime following an earthquake would be disastrous. Data centres, TV studios and even newspaper print facilities require continuous operation. In some buildings, the focus of seismic design is protection of the building contents, possibly far more valuable than the building itself. Laboratories, libraries and semiconductor plants are examples. Some owners desire to minimize damage and damage repair costs. Perhaps their investment in seismic isolation is an alternative to paying earthquake insurance premiums and allowing early, if not immediate, re-occupancy of a building following an earthquake. Government, commercial and even retail buildings such as shopping centres benefit from this approach. There are also many buildings of historic significance that have been seismically isolated to protect their fabric from damage and preserve their architectural qualities. These buildings use seismic isolation as their primary structural strategy. Several case-studies are presented in Chapter 7. The final reason for implementing seismic isolation is occupant safety. Although modern conventionally designed buildings provide a high degree of seismic safety, occupants can be even safer and certainly experience far less trauma during an earthquake in a seismically isolated building. This provision of increased personal safety is evidenced in growing numbers of seismically isolated apartment buildings or condominiums and houses, particularly in Japan.

Seismic isolation has been applied to both new and existing buildings. Certainly, the technology is introduced more easily into a new building, but there are now hundreds of

2.12 Oakland City Hall, Oakland, USA, retrofitted 1995

buildings around the world retrofitted by seismic isolation. They include a wide diversity of both heritage and newer buildings. As well as preserving history in the form of iconic historic buildings, both building occupants and business continuity are being protected in modern buildings. In each case the owner is convinced that seismic isolation will achieve one or more of the six objectives of seismic isolation listed above.

This brief overview of the history and diversity of seismically isolated buildings also needs to acknowledge the seismic isolation of just parts or contents of buildings, rather than buildings themselves. Many examples may be found where sensitive or fragile equipment is isolated. Large specialized industrial equipment to smaller computer racks and server cabinets can all be mounted on their own isolation devices. Even individual sculptures are isolated within conventionally constructed museums. It appears that there are few, if any, types of high-value buildings, equipment or contents that cannot benefit from seismic isolation.

2.13 Museum of New Zealand – Te Papa Tongarewa, Wellington, 1998

2.14 Wellington Regional Hospital, New Zealand, 2008

Requirements of seismic isolation

THE previous chapter highlighted the conceptual simplicity of seismic isolation by likening it to forming a slippery plane between the foundations of a building and its superstructure. However, while the capability for horizontal movement is the first requirement for an isolated structure, there are four others as well: vertical support, re-centring, restraint and damping. Contemporary seismic isolation systems incorporate each of these requirements.

2.15 The Indumotora building, Santiago, Chile, is a commercial building

Source: Sabbagh Architects

Movement capability

The superstructure of a building needs to be isolated from the violent to-and-fro horizontal shaking of the ground by providing for relative movement between superstructure and ground. Ideally, the ground and the foundations of an isolated building vibrate back-and-forth in any direction without transferring kinetic energy into the building. When the ground moves, the superstructure should remain motionless. In practice it is more complex, but movement capability is provided by elastomeric bearings from natural or artificial rubber, or sliding surfaces of Teflon and stainless steel, and by complete horizontal separation of the superstructure from the ground.

Vertical support

In order to allow almost unrestrained horizontal movement, an isolation system must support the weight of the building. Hardware, such as most types of bearings, not only provides

2.16 Innovation Centre UC, Santiago, Chile

Source: Innovation Center UC-Rodolfo Jara Verdugo

2.17 Ñuñoa Capital Building, Santiago, Chile. An apartment and office building

Source: Ñuñoa Capital-Inmobiliaria Armas

2.18 The TV house, Suita, Osaka, Japan

Source: Noriyoshi Morimura

2.19 TV house detail of bearing and services connection to house

Source: Noriyoshi Morimura

2.20 Bahá'í Temple, Santiago, Chile

Source: Board of the Bahá'í Temple, South America

movement capability but transfers the entire weight of a building to its foundations. Bearings are normally located under columns and structural walls. They raise the superstructure above the ground and define discrete points where gravity loads are transferred into the foundations.

Re-centring

During earthquake shaking, which is random in extent and direction, a superstructure tends to drift in a particular direction. Depending on the intensity and duration of an earthquake, a superstructure could move further than the movement capability of the bearings, damaging them. A re-centring or restoring force is always provided. It keeps bringing the superstructure back to its original position. It prevents bearings incrementally having their movement capacity reduced by long-duration shaking.

Restraint

In theory, to achieve the maximum degree of seismic isolation, a superstructure should bear on a frictionless surface. Unfortunately, horizontal forces other than from earthquakes also can move buildings sideways. Wind is the most obvious example. A seismic isolation system must prevent motion along the plane of isolation during wind gusts. Not only could the building move excessively over its bearings but building occupants could suffer from motion sickness. Such a 'windblown' building would, no doubt, also attract unwanted media attention! Designers must therefore provide isolation systems with minimum horizontal restraint to withstand code wind forces.

Damping

Finally, damping is necessary. Just as shock absorbers prevent intolerable vertical vibrations when a car rides over bumps, damping reduces the overall dynamic response of a building. It lessens the relative horizontal movement between the superstructure and foundations. Bearings and other details can then be designed for far less movement, reducing both cost and the width of gaps that accommodate movement.

Reduction of earthquake response

HAVING identified the five requirements of a seismic isolation system we now introduce the two principles behind the effectiveness of seismic isolation – period-shift and damping.

First, it is necessary to appreciate the dynamic characteristics of earthquakes. Usually, most of their dynamic energy is contained in high frequency vibrations, say between one and ten cycles per second. These vibrations are very effective in causing dynamic resonance in buildings that naturally vibrate horizontally at the same frequencies. Every building, when jolted suddenly, vibrates at its own natural frequency. Or, expressed more commonly, particularly by structural engineers, every building has its own natural period of vibration. The natural period of a building is the time taken for a complete cycle of vibration, the inverse of the natural frequency. A building with a natural period of 0.1 seconds has a natural frequency of ten cycles per second.

For every earthquake, we can plot a graph of the horizontal acceleration response of buildings with increasing natural periods (Figure 2.21). Such a graph is termed a 'response spectrum'. Response spectra from many earthquakes are plotted, averaged and simplified before being introduced into countries seismic design codes. Although the shapes of response spectra vary for each earthquake, site soil conditions and seismic code, they all display the common feature that for periods longer than about 0.5 seconds, the acceleration response of the building diminishes. Seismic isolation exploits this fact – that a building with a longer period of vibration (a more flexible building) experiences less shaking than a shorter period building. Compared to an otherwise identical conventional building, an isolated building is allowed to move freely but to re-centre through horizontal spring action. Seismic isolation therefore creates a very flexible building and introduces a 'period-shift'. A period shift, often between two to three seconds, dramatically reduces acceleration response. Period-shift is the main reason for the effectiveness of seismic isolation in reducing superstructure accelerations that otherwise cause widespread damage.

Second, and as mentioned previously, damping also reduces the dynamic response of a building (Figure 2.22). Damping reduces resonance. It causes the amplitude of vibrations to decay by absorbing dynamic energy. Damping in buildings is expressed as a percentage of critical damping. Rather than a building vibrating to-and-fro after being suddenly pushed sideways, a critically damped system just returns to its original position. Every building has some inherent damping depending on its structural materials. For example, reinforced concrete buildings possess greater damping than those with steel framing. But seismic isolation systems typically provide three to five times more damping at the isolation interface.

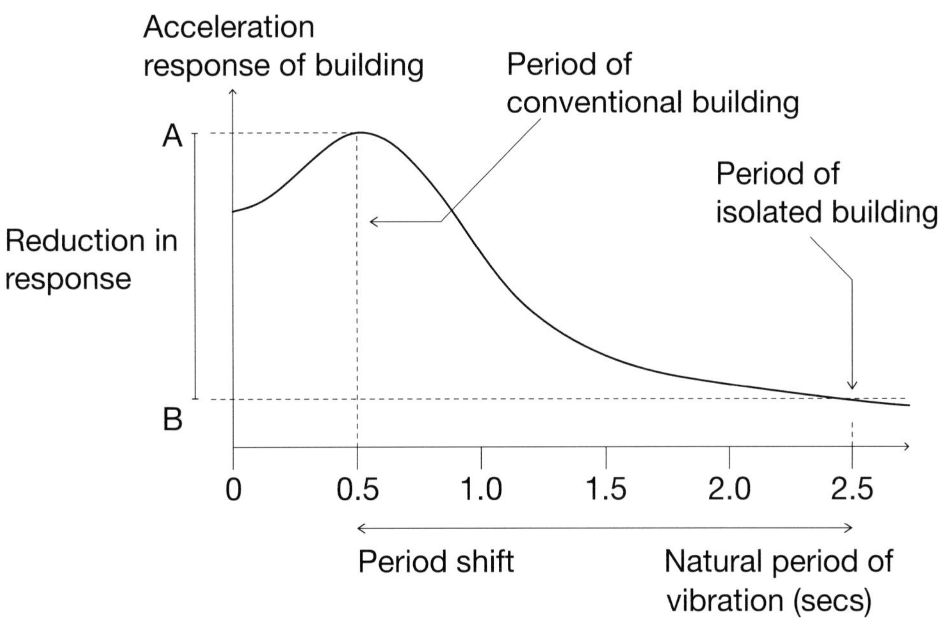

2.21 A response spectrum shows how the acceleration response of a building to a typical earthquake varies according to the natural period of the building. The effectiveness of seismic isolation is due largely to the 'period-shift'

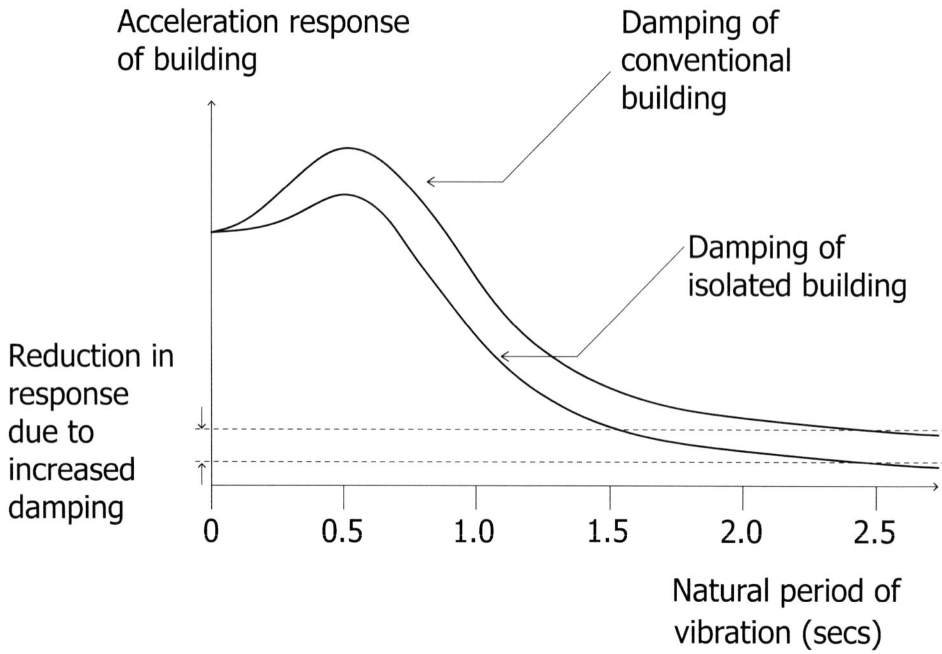

2.22 A response spectrum of a typical earthquake showing the reduction in the response of an isolated building due to increased damping

Generic isolation hardware

EACH isolation system must satisfy the five requirements of movement capability, vertical support, re-centring, restraint and damping. Typical seismic isolation hardware that is described in more detail in Chapter 3 meets two or more of these requirements (Figure 2.23). Lead-rubber bearings, for example, meet all five. Their height enables them to move horizontally. When that occurs, the elasticity of their rubber causes them to spring back or re-centre. Their steel plate-rubber sandwich construction enables them to support the building weight without excessive vertical settlement, and the lead plug both restrains wind movement and dampens vibrations by absorbing energy.

Curved sliders also integrate all five requirements, but via totally different mechanisms. Movement is provided by a Teflon-coated slider that bears on highly polished stainless steel surfaces. Gravity loads pass through the slider and to the lower stainless steel plate. The static friction along this interface is sufficient to provide wind restraint, and then when exceeded in an earthquake, dynamic friction is the method by which energy is absorbed and the system damped. The concave curved surface means that whenever the slider displaces horizontally it also rises. The weight from the building constantly drives the slider down the curved surface to cause the re-centring action.

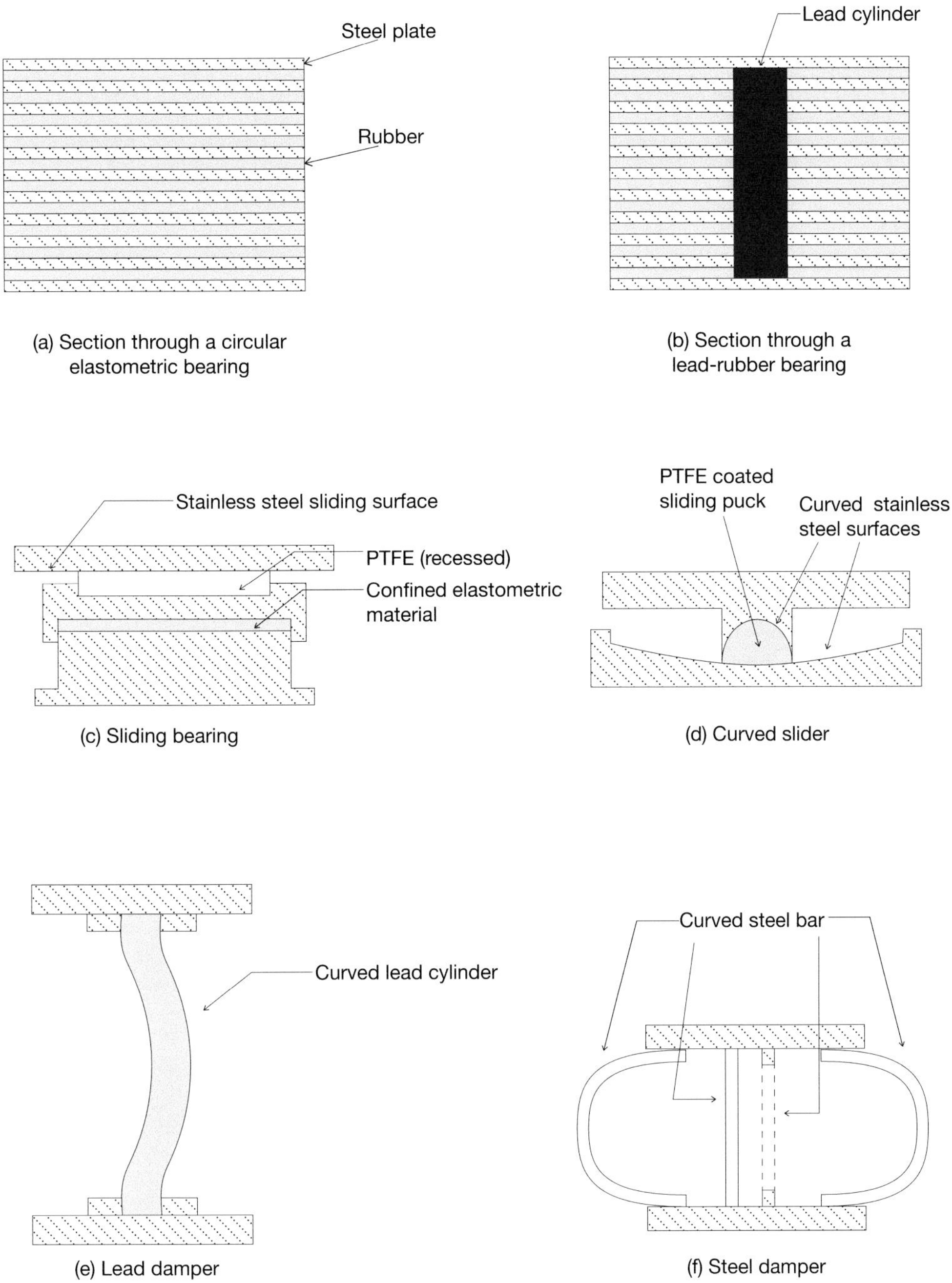

2.23 Different types of seismic isolation hardware: (a) Rubber bearing, (b) Lead-rubber bearing, (c) Slider, (d) Curved slider, (e) Lead damper, and (f) Steel damper

Horizontal sliding bearings provide movement capability and support the building weight. Their internal friction provides restraint and damping. However they need to be supplemented by other spring-like devices to provide re-centring.

Notes

1 Naderzadeh, A., 2009, 'Application of seismic base isolation technology in Iran', *Menshin* (Japan), no. 63, pp. 40–7, p. 42.

2 Christopoulos, C. & Filiatrault, A., 2006, *Principles of passive supplemental damping and seismic isolation*, Pavia: IUSS Press.

3 Pan, P., Zamfirescu, D., Nakashima, M. *et al.*, 2005. 'Base-isolation design practice in Japan: introduction to the post-Kobe approach', *Journal of Earthquake Engineering*, vol. 9, no.1, pp. 147–71.

4 Martelli, A., Forni, M. & Clemente, P., 2012, 'Recent worldwide application of seismic isolation and energy dissipation and conditions for their correct use', Proceedings of the 15th World Conference on Earthquake Engineering, Lisbon.

5 Zhou, F.L., Tan, P., Heisa, W.L. *et al.*, 2013, 'Lu Shan earthquake M 7.0 on 2012.4.12 and recent development on seismic isolation, energy dissipation & structural control in China', Proceedings of the 13th World Conference on Seismic Isolation, Energy Dissipation and Active Vibration Control of Structures, Sendai.

– 3 –

Seismic isolation systems and hardware

Introduction

FOR most of its brief history, seismic isolation can be considered a passive system. That is, the hardware is installed in buildings and sits there passively, waiting to be activated by an earthquake. Even during earthquake shaking, isolation devices passively endure the dynamic movement while simultaneously performing their damage mitigation and lifesaving functions. They absorb and dissipate earthquake energy, dampening down movements. They provide the horizontal flexibility which is at the heart of seismic isolation, all the while bearing the weight of the superstructure above. And finally, their re-centring forces are always bringing the building back to its original position above the foundations. Once installed, protected against fire if necessary, and subject to regular inspection and maintenance, isolation devices are left alone to passively play the roles outlined above that they are designed for.

However, recently, research and development has focussed upon active systems. Usually relying on movement sensors and computer-controlled hydraulic rams, active systems counteract the ground movement. The ultimate goal of active seismic isolation is to keep a building stationary while everything around and under it shakes (Figure 3.1). As you would expect, active systems are high-tech and very expensive. Although they control wind vibrations in several tall buildings, apart from a demonstration building discussed near the end of this chapter, they have yet to find acceptance. But as individuals and companies become both become more risk-adverse and enamoured by high-tech solutions, their uptake is sure to increase in the future, particularly shortly after damaging quakes.

Since active systems can be considered still under development, the focus of this chapter is upon passive systems. We discuss the three categories of systems – elastomeric and dampers, sliding, and other systems in the following sections.

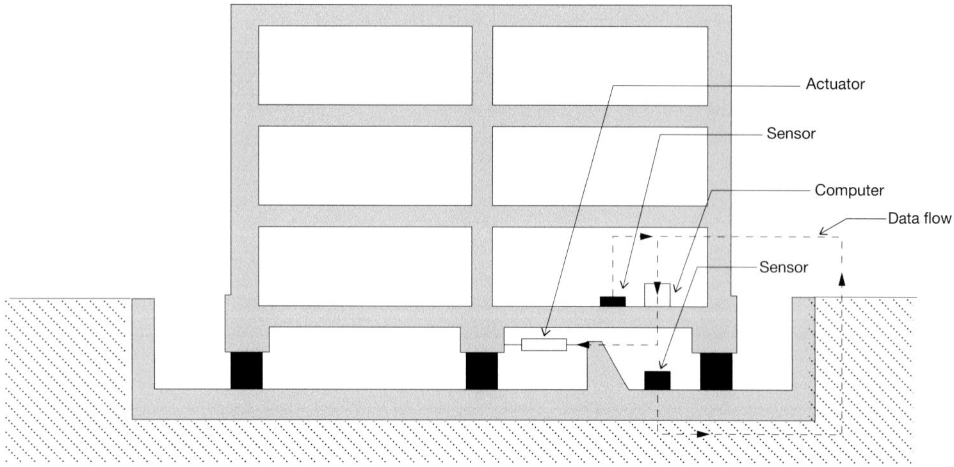

3.1 An active isolation system keeps the building stationary during shaking. Sensors feed information to a computer-controlled hydraulic ram

Elastomeric systems and dampers

THE most basic component of these systems is the elastomeric bearing. Used in bridges for over 100 years, these bearings allow decks to move over piers and abutments during thermal expansion and contraction. The main difference between bridge bearings and those providing seismic isolation in buildings (and bridges) is their height. Seismic movements, which are typically ten times greater than thermal movements, require a greater thickness of rubber.

Elastomeric bearings consist of alternating layers of usually natural rubber and thin steel plates, steel shims. The shims prevent excessive lateral bulging and settlement under compression (Figure 3.2). The bearings range in diameters from 500 mm to 1500 mm depending on the compression force acting and the horizontal design movement. The approximate thicknesses of the rubber and steel shims are 10 mm and 3 mm respectively. The bearings are therefore very stiff vertically yet flexible for horizontal movement. Typically a bearing might settle up to 1–3 mm initially due to the column load acting upon it and a further 50 per cent over the following decades. Yet it can move horizontally 300–500 mm. These bearings are extremely robust. The horizontal rubber layers and protective exterior layer are vulcanised to the steel shims. Under long-term loads they are stable. As the rubber ages it hardens slightly, but a bearing is not detrimentally affected by oxidation.[1]

Elastomeric bearings placed under the columns of a seismically isolated building are well-suited for supporting the building weight, providing horizontal flexibility and the necessary re-centring force. However they may lack significant damping and might be too flexible under wind gusts. Three approaches are taken to provide additional damping.

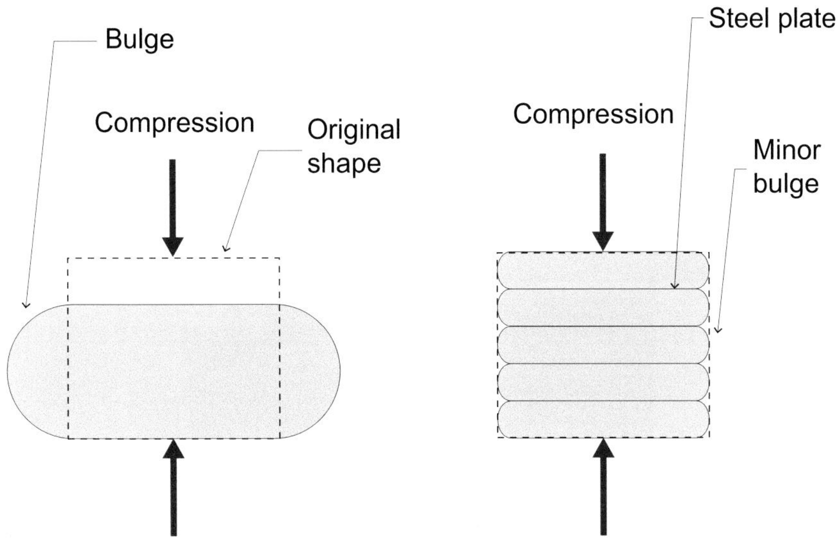

3.2 The effectiveness of steel shims at reducing the settlement of an elastomeric bearing

The first is to insert one or more cylindrical lead plugs of approximately 60–150 mm diameter into the bearing. This is the widely used lead-rubber bearing. The lead provides rigidity against wind and dissipates large amounts of energy as it yields and recrystallizes during an earthquake (Figures 3.3 and 3.4). During their development, these bearings were subject to punishing test regimes to ensure satisfactory short and long-term performance. In most installations two prototypes must survive testing so severe as to disqualify them for eventual incorporation in the building.

Additional or alternative damping can also be provided by adjusting the properties of natural rubber. Proprietary compounds are added and the curing process adjusted, or synthetic rubber is used. High-damping rubber bearings are commonly used as seismic isolators.

Provision of damping in conjunction with elastomeric bearings is also achieved with supplementary external dampers. Lead dampers are particularly popular in Japan. These pure lead curved cylinders are typically 200 mm in diameter and about 1 m high. They are thoroughly tested and accept maximum displacements of up to 800 mm.[2] A less common type of lead damper is the lead-extrusion damper. It is like a scaled-up car shock absorber but filled with lead rather than oil (Figure 3.5). The damper connects a superstructure to its foundations, across the isolation plane. When the ground moves, exceeding the yield force of the damper, the lead against the bulge of the shaft melts and then recrystallizes after the shaft has moved through the lead.[3] Since the damper only works in the direction of its length, at least four dampers, two in each orthogonal direction and widely separated in plan are required in any building.

3.3 A schematic view of a lead-rubber bearing

Source: FIP Industriale S.p.A., Italy

The Wellington Central Police Station incorporates lead-extrusion dampers (Figure 3.6). The building has an isolation system based on long piles in sleeves. Firm bearing material was overlain by relatively soft soils so piles were inevitable and sleeves were placed around them to separate the piles from the ground in order to provide horizontal flexibility. The bases of some piles are socketed deeply into rock. These piles function as vertical cantilevers and provide the re-centring force. Damping is provided by lead-extrusion dampers located around the perimeter of the building in the basement (Figures 3.7 and 3.8).

Steel dampers also provide damping. There are many variants, but one type used in Japan is shown in Figure 3.9. Devices such as this endure numerous large horizontal displacements

3.4 A lead-rubber bearing in a building basement. Note the small compression bulges in the rubber between the steel shims

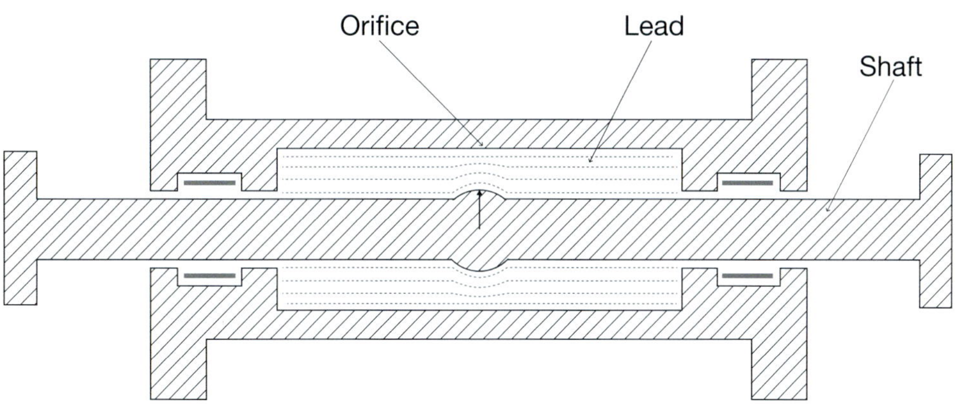

3.5 A schematic view of a lead-extrusion damper

3.6 Wellington Central Police Station, New Zealand, 1991

3.7 A lead-extrusion damper in the basement of the Wellington Central Police Station

without breaking. They dissipate energy by their steel members yielding. At Union House, Auckland, which is founded on long sleeved piles like the Wellington Central Police Station, horizontal restraint and damping is provided by mild steel tapered dampers (Figures 3.10 and 3.11).

Hydraulic dampers filled with oil or a silicon fluid are another option. They are less frequently used than other dampers, possibly due to their cost, although in theory they are the most effective. Their advantage is in reducing the lateral loads for which the superstructure need be designed. This is because the force they transfer to the superstructure is proportional to the velocity of movement. When there is little force in a nearly centred elastomeric bearing, the oil damper will be moving fast and exerting its maximum force. Then, when the elastomeric bearing is nearing the end of its displacement, subject to maximum force and the velocity is low, the hydraulic damping force has died away.

Sliding devices

Two types of sliding devices are common in seismic isolation systems. First, relatively simple sliding bearings, and second, the far more sophisticated curved sliders that, like lead-rubber bearings, comprise a complete isolation system themselves.

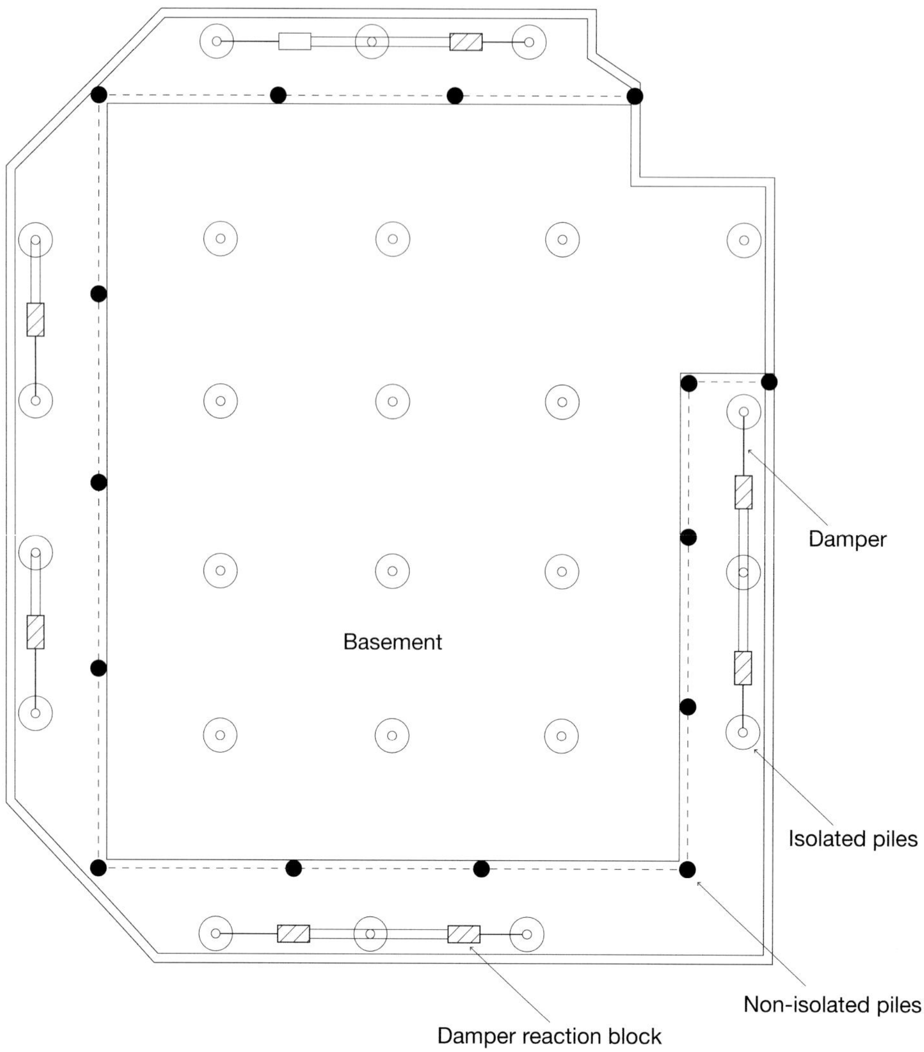

3.8 A basement plan showing the location of lead-extrusion dampers, Wellington Central Police Station

Sliding bearings transfer compression load and allow horizontal sliding. In most cases the sliding surfaces in contact are stainless steel and polytetrafluoroethylene (PTFE), commonly known as Teflon. These types of bearings have been used for many years to accommodate thermal movements in bridges. They are known as pot bearings due to the way a thin layer of rubber is confined in a pot-shaped ring (refer to Figure 2.23(c)). The rubber allows the sliding surfaces a small degree of rotation, and is confined to allow the bearing to carry high compression loads. These sliding bearings are often used in seismic isolation and are available from many manufacturers.

3.9 A steel damper integrated with an elastomeric bearing

Source: Nippon Steel & Sumikin Engineers

The main requirement of a sliding bearing is that it slides freely with a minimum of friction. Although stainless steel and Teflon tend to stick together before being released by horizontal movement, very low coefficients of dynamic friction in the order of one per cent are achievable for low velocities, rising to a more typical and constant value of approximately 10 per cent at the higher velocities experienced during earthquakes. Higher friction values can provide a valued source of additional damping. In contrast to conventional construction materials, low coefficient of friction values occur when the compression stresses are greatest. It is very important to keep the sliding surfaces clean.

Sliding 'pot-type' bearings are unsuited for seismic isolation by themselves due to their lack of any re-centring capability. They are often used in conjunction with natural rubber bearings or lead-rubber bearings. The relative proportions of elastomeric and sliding bearings vary from project to project in order to optimise dynamic performance and reduce costs. Sliding bearings, with their high compression load capacity, are often placed under the most heavily laden columns or structural walls.

A significant advance in the use of sliding bearings for seismic isolation occurred in the late 1980s when flat stainless steel surfaces were curved into concave forms. Now the

3.10 Union House, Auckland, New Zealand, 1983. Superstructure wind and seismic loads are resisted by the exterior bracing system

3.11 Tapered steel damper that connects the superstructure to the foundation provides damping when it yields during an earthquake

sliding bearing possessed re-centring capability, and along with damping provided by friction between its sliding surfaces, it met all the requirements for seismic isolation. As the sliders that support the building move along the concave surfaces the building rises. Gravity loads from the weight of the building above force sliding towards the centre of the bearing thereby causing re-centring (Figure 3.12).

The first friction pendulum bearing was applied in a 1994 retrofit project, and since then the original design has been developed from a single pendulum operation with two sliding surfaces to the Triple Pendulum™ with its four surfaces (Figure 3.13). In this latest version of what are generically referred to as curved sliders, the radii of curvature of the sliding surfaces, which determine the natural period of the bearings, vary and the surfaces can possess different friction values. This improves the dynamic performance over a wide range of earthquake motions and allows smaller diameter bearings. Another advantage of curved sliders is their thinness (Figure 3.14). This means more compact construction detailing and a reduction of forces acting on structural members adjacent to the bearings. Like elastomeric systems these bearings have been, and in every project are, subjected to strenuous full-scale testing regimes.

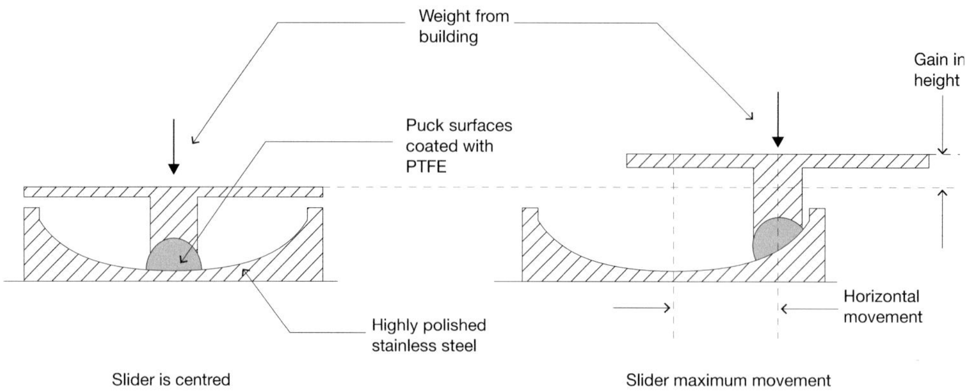

3.12 Diagrammatic representation of a Friction Pendulum Isolator™. As the ground moves to the left, the puck slides along and up the concave surface. Friction between the sliding surfaces provides damping and the building weight causes the puck to slide back down the slope, or re-centre

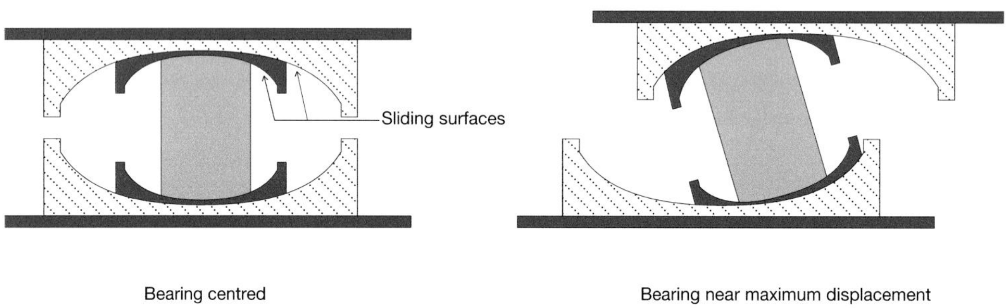

3.13 Schematic view of a Triple Pendulum™ bearing centred, and then at its maximum displacement

Other systems and devices

While the seismic isolation systems and devices discussed above constitute the vast majority of both those used currently and in the past, it is worth describing some that are less common as they may be applicable in certain situations. Although their configuration is very different from what has already been presented, they all encapsulate the basic principles of seismic isolation, namely increasing flexibility (the natural period of vibration), allowing large displacements along the isolation plane and providing increased damping.

The Sendai Mediatheque, Japan, designed by Toyo Ito, exploits the flexibility of its sub-surface structure (Figures 3.15 and 3.16). To achieve the necessary delicacy of its iconic braced-tube clusters yet protect them from seismic damage, some method of isolation was

3.14 A cut-away view of a curved slider

Source: FIP Industriale S.p.A., Italy

required. In this case, the interface of superstructure and foundations consists of four basement ductile moment frames.[4] The three beams per frame are designed to yield in a large earthquake (Figure 3.17). As sacrificial elements, they suffer structural damage while absorbing earthquake energy. In the process they reduce and limit the forces transferred into the superstructure. This bespoke isolation system does not provide the large period shift of typical isolation systems. Perhaps this was one of the reasons for suspended ceilings collapsing during the 2011 Tōhoku earthquake.[5, 6]

Another form of isolation can be achieved by letting structure, such as a structural wall, rock to-and-fro. Once a structure starts rocking during an earthquake its natural period increases substantially, partially isolating it from the earthquake shaking.[7] Damping occurs as the ends of the wall impacts the soil, but in some cases additional damping devices are provided. Although allowing the structure to rock may protect it and save foundation costs,

3.15 Sendai Mediatheque, Sendai, Japan

3.16 One of four braced hollow tubes that resist most of the seismic loads acting on the building

some of the benefits of a full seismic isolation system, such as reduction of damage to contents, will not be realized.

The Hermès building, Tokyo, designed by Renzo Piano Building Workshop, is a high profile example of seismically isolating a building by allowing rocking (Figure 3.18). An article elaborates:

> At 50 m tall and with a main structural span of only 3.8 m, the unusual slenderness of the structure results in high overturning moments during an earthquake and high levels of tension in the columns. The engineer . . . found inspiration in the tall, thin wooden Buddhist pagodas of Japan . . . the same principle was adopted, with the columns on one side of the frame being held in base joints that allow uplift and rotation simultaneously and seismic energy to be absorbed by viscoelastic dampers.[8]

In all the examples of seismically isolated buildings considered to date, isolation planes are located at the bases of the buildings. However, in some situations it is advantageous for the isolation plane to be located in a middle storey. An example of this approach is found at

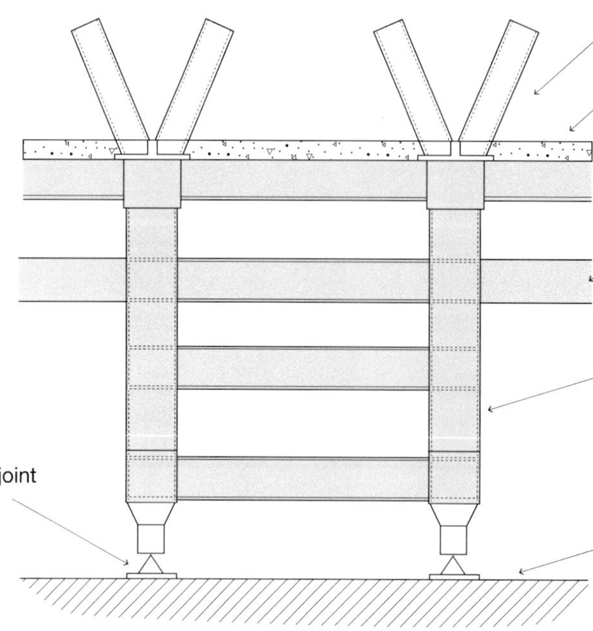

3.17 An elevation of a basement ductile steel moment frame that protects the superstructure

the Iidabashi First Building, Tokyo.[9] Eight floors of offices that rise above the ground floor retail area provide generous (16 m) column-free spaces. On top of this structure, and seismically isolated from it by rubber bearings and lead dampers, sits a further five levels of apartments. When the main structure moves during an earthquake, the isolated upper floors vibrate out-of-phase. The levels of force within the main structure are thereby reduced. The isolation level accommodates building services, and the two external apartment elevator towers that cantilever above the main structure are seismically separated from the apartments but linked to them by seismic movement joints.

There are also many less common proprietary systems, often developed for special situations. The RoGlider™ is an example (Figure 3.19). Suitable for both light and heavy vertical loads, a puck coated with PTFE slides between two stainless steel plates. Rubber membranes that connect the puck to the plates provide the re-centring force. At the Wanganui Hospital these bearings were designed for a maximum displacement of 450 mm and their 10 per cent coefficient of friction provides restraint against wind load as well as dynamic damping. Manufacturers' catalogues can be consulted for other examples.

Researchers around the world are continuing to develop new systems. Every year, one or two new ideas are presented in professional journals or at conferences. Recently, two

3.18 Hermès building, Tokyo, Japan, 2001. The structure is designed to partially self-isolate by rocking

3.19 A RoGlider™ installed between a foundation pad and substructure framing under the Wanganui Hospital, New Zealand

researchers proposed telescopic columns as a means of isolating high-rise buildings.[10] Unlike any other columns, these would contain many small yielding steel plates that would absorb seismic energy when compressed and extended. It is too soon to tell if this and many other innovative approaches will eventually be observed on building sites.

Active systems

ACTIVE isolation systems are used in small numbers of very tall buildings, mainly to control wind vibrations. They are similar to, but more sophisticated versions of, passive tuned-mass dampers, designed to vibrate out-of-phase with the superstructure and hence reduce overall dynamic response. In active systems, sensors detect movement of the building and the ground, and signals are sent to computer-controlled hydraulic jacks. The jacks force a heavy mass in the building, or the whole building, back-and-forth with similar frequencies to the earthquake shaking to keep the building as stationary as possible.

Obayashi Corporation's Technical Research Institute in Tokyo is the first building to use this technology (Figures 3.20 and 3.21). The system, designed to reduce building shaking

3.20 Technical Research Institute, Obayashi Corporation, Tokyo, Japan. An active isolation system protects the building

3.21 One of the computer-controlled hydraulic jacks

to 1/30th of actual ground shaking has required very sophisticated design. All known eventualities have been taken into account. In the event of an electrical power failure the hydraulic jacks are powered from tens of high pressure oil cylinders, and in the worst case scenario, the building relies upon a back-up passive elastomeric isolation system. Needless to say, this system is extremely expensive, but the manufacturer is trying to reduce costs for implementation in hospitals and manufacturing premises, such as for precision measuring equipment. There is also an option of applying this technology to just one floor of a building.

Notes

1 Kelly, T.E., Skinner, R.I. & Robinson, W.H., 2010, *Seismic isolation for designers and structural engineers*, Kanpur: NICEE, p. 93.

2 Pan, P., Zamfirescu, D., Nakashima, M. *et al.*, 2005, 'Base-isolation design practice in Japan: introduction to the post-Kobe approach', *Journal of Earthquake Engineering*, vol. 9, no. 1, 147–71.

3 Kelly, T.E. *et al.*, p. 57.

4 Sasaki, M., 2002, 'Sendai Mediatheque, Japan', *Structural Engineering International*, no. 3, pp. 146–8.

5 'March 11 Earthquake at the sendai mediatheque.' Available from: www.youtube.com/watch?v=TKgURstRt_A (accessed: 21 January 2016).

6 'Sendai mediatheque's damages of the Great East Japan Earthquake.' Available from: www.youtube.com/watch?v=0D4V803Z5O8 (accessed: 21 January 2016).

7 Rezai, M., Patterson, A. & Hubick, G., 2012, 'Nonlinear seismic analysis and retrofit of BC Place Stadium using rocking foundation and viscous dampers', Proceedings of the 15th World Conference on Earthquake Engineering, Lisbon.

8 Heneghan, T., 2001, 'Japanese lantern', *The Architectural Review*, vol. 220, no. 1255, pp. 78–81, p. 81.

9 Tasaka, M., Mori, N., Yamamoto, H. *et al.*, 2008, 'Applying seismic isolation to buildings in Japan – retrofitting and middle-story isolation', Proceedings of the 18th Analysis and Computation Speciality Conference, ASCE, Vancouver, p. 6.

10 Hosseini, M. & Farsangi, E.N., 2012, 'Telescopic columns as a new base isolation system for vibration control of high-rise buildings', *Earthquakes and Structures*, vol. 3, no. 6, pp. 853–67.

– 4 –

Effectiveness of seismic isolation

Introduction

When modern seismic isolation was first applied to buildings in the 1980s, its claim for effectiveness was based entirely upon computer modelling and laboratory testing of isolation hardware, such as lead-rubber bearings. Those early applications of the then new technology required a certain degree of faith in the theory of seismic isolation as well as in the computer analyses that constituted the basis of the structural engineering designs.

Now, in 2016, the effectiveness of seismic isolation not only continues to be validated in virtual environments, but it has been demonstrated both in the real world of full-scale testing and, more significantly, in post-earthquake urban centres of several countries. Seismically isolated buildings have now been tested by actual earthquakes ranging from those of low intensities to more intense motions similar to design earthquakes. These buildings protected by seismic isolation are on sites characterized by different soil conditions, in particular soft soils, and have been subject to damaging earthquakes from nearby faults, as well as those several hundreds of kilometres away.

The following sections demonstrate the effectiveness of seismic isolation as determined by three quite different approaches.

Computer modelling

The effectiveness of seismic isolation is confirmed each time an isolated building is designed. In most such designs structural engineers subject a computer model of the structure, including the isolation system, to several different earthquake records in order to simulate the strong shaking expected in the design earthquake. Often, these models are complex. They attempt, as accurately as possible, to predict the dynamic response of the isolated structure (Figure 4.1). Then, once the main design variables, such as the maximum displacement across the isolation plane and the maximum forces acting within the superstructure, have been determined, detailed design can be completed.

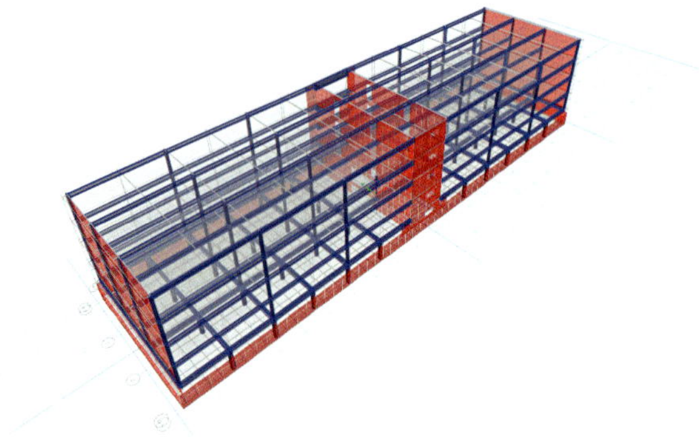

4.1 A computer model of the IHEM research building, Mendoza, Argentina

Source: Agustín Reboredo

As well as the effectiveness of seismic isolation being proven during every building design, our initial and on-going confidence in seismic isolation is also grounded in the large body of research that began in the 1980s. Since then, tens of PhD theses have explored aspects of the topic, hundreds of research papers presented at conferences around the world and numerous scholarly books written. Each of these publications, to some extent, is based on the results of computer modelling. For example, a single table within one book summarizes findings from 100 computer analyses.[1] Research is on-going. At earthquake engineering conferences around the world, we can guarantee that new research findings on seismic isolation are being presented and discussed.

Physical testing

It is comforting that the effectiveness of seismic isolation is not only reliant upon theoretical and virtual studies. Physical testing has been a vital component in the early development of seismic isolation, and in its on-going quality assurance programme. Initially, just devices like lead-rubber bearings, curved sliders and many types of dampers were tested on their own. These relatively small test set-ups were and still are challenging enough. They have to replicate the large gravity forces acting on bearings plus simulate the dynamic motions that occur during large earthquakes (Figures 4.2 and 4.3).

With the recent availability of hydraulically activated shaking tables, physical testing can be scaled up. Now, not only can several isolation devices be tested simultaneously, but entire isolated buildings experience recorded strong motion earthquake records. For example, the largest shaking table in the world, E-Defense, subjects concrete and steel buildings up to five storeys high to 3-D shaking. In one test of a 500-tonne steel frame building, the effectiveness

4.2 The testing of elastomeric bearings. Two bearings for the Hospital of Light, Lisbon being tested

Source: FIP Industriale S.p.A., Italy

4.3 A bearing for the China Basin Landing office building, San Francisco being tested

Source: Dynamic Isolation Systems Inc.

of two isolation systems were compared and confirmed when compared to a conventional fixed-base building: 'Both isolation systems [elastomeric and curved slider] successfully mitigated damage in the superstructure from a range of strong ground motions',[2] with lesser measured floor accelerations than in the fixed-base case. Due to their different isolation periods, the elastomeric system more strongly attenuated floor accelerations at lesser ground acceleration, whereas the curved slider was more effective at higher ground accelerations. The three test earthquake ground motions simulated 'frequent', 'long-duration subduction' and 'near-fault' motions; the last two of which are well recognised as providing the most severe test of a seismically isolated system.

In another test, also at E-Defense, the earthquake performance of an isolated and a fixed-base four-storey concrete building were investigated.[3] The aims of this test were to check the effectiveness of seismic isolation for both near-fault and long period motions that are expected on sites underlain by soft soils that vibrate slowly. The focus was on preventing damage to furniture and mobile equipment. Seismic isolation proved to be very effective for the near-fault earthquake motion. Superstructure accelerations were only 32 per cent of the ground accelerations while the fixed-base superstructure amplified ground accelerations (measured at roof level) by 3.6 times. Therefore, the maximum levels of acceleration in the seismically isolated superstructure were 1/10th of those in the fixed-base building, thereby providing far greater protection to building contents. However, when the same buildings were subject to extreme long-period earthquake vibrations the effectiveness of seismic isolation diminished greatly. The long-period nature of the shaking resonated the flexible isolation system. Although the isolation system was unable to reduce the level of ground accelerations transferred into the superstructure, the maximum acceleration of the isolated building was only 66 per cent of that in the fixed-base building.

This test illustrated the need for castors under mobile equipment to be locked whenever stationary. But more significantly, it emphasises how the effectiveness of seismic isolation depends on the dynamic characteristics of the earthquake waves, they themselves dependent upon the proximity of the epicentre and the softness and depth of underlying soils. This is why geotechnical and seismological advice must always be sought before deciding to adopt seismic isolation.

Measurements and observations during earthquakes

EVEN though the effectiveness of seismic isolation is demonstrated extensively through computer modelling and physical testing there is no substitute for measurements and observations during and after damaging earthquakes. No sceptic is satisfied by only the previous two methods that demonstrate effectiveness.

Demonstration of effectiveness in a real earthquake is required. The ultimate test would consist of isolated buildings being subject to shaking strong enough to cause maximum design displacements at their planes of isolation. While one day this will occur, currently we have to be satisfied with observed performances in lower intensity yet still significant strong ground motions.

Three approaches can be used to evaluate the effectiveness of seismic isolation after an earthquake. We can compare the behaviour of identical side-by-side isolated and conventional buildings, or situations where non-identical buildings are in close proximity. We can also make use of acceleration measurements taken beneath and within individual isolated buildings to gauge the effectiveness of seismic isolation. Finally, we can make general comparisons between the performance of earthquake-affected isolated buildings and conventional buildings.

Identical side-by-side isolated and conventional buildings are uncommon but are found in Chile and Japan. Moroni *et al.* (2012) report on two four-storey reinforced concrete confined masonry buildings in Santiago, one of which is seismically isolated. The buildings were affected by the 27 February 2010 M_w 8.8 Maule earthquake.[4] Due to the large epicentral distance and low intensity ground motion at the site, the isolated building moved 80 mm along its isolation plane. Its roof acceleration was less than the maximum ground acceleration as compared to the conventional building whose roof experienced a four-fold amplification.

Two identical and instrumented buildings have been built at Tōhoku University, Japan (Figure 4.4). During the 2011 Tōhoku earthquake, the isolation plane movement was also 80 mm. As in Santiago, the isolated building experienced less acceleration than the ground and its neighbour, but in this case only by a factor of 2.0.[5]

4.4 Two identical buildings, one base-isolated and the other a conventional fixed-base building, at Tōhoku University, Sendai, Japan

Source: Shimizu Corporation & Tōhoku University

Several comparisons have been made between similar seismically isolated and conventional buildings located within the same area. Ron Mayes (2007) cites two examples.[6] First, the seismically isolated Japanese Postal Computer Center suffered no damage during the devastating 1995 M_w 7.1 Kobe earthquake. Instrumentation showed the isolation system reduced ground accelerations by a factor of 3.4. A similar instrumented six-storey reinforced concrete building in the next block amplified the ground shaking by a factor of 3.0 at the roof. So compared to the conventional building, isolation reduced the shaking at roof level by a factor of 10. A similarly impressive case is cited following the 1989 Loma Prieta earthquake, California: 'The University of Southern California (USC) Hospital – the world's first base-isolated hospital (Figure 4.5) – had no damage at all, while the Los Angeles Country General Hospital complex 1 km away from the isolated USC hospital suffered $389 million in damage.' The benefits of seismic isolation also became apparent in the Rikkyo University Chapel building, Tokyo during the 2011 Tōhoku earthquake.[7] This historic brick masonry building had been seismically retrofitted by a seismic isolation scheme. During the earthquake its bearings and dampers moved 40 mm, and the superstructure experienced only 38 per cent of the peak ground acceleration. Several metres away an instrument on the second floor of a conventional brick building indicated an amplification of the ground accelerations by a factor of 3.0.

4.5 USC Hospital, USA, the world's first seismically isolated hospital, survived the 1989 Loma Prieta earthquake without damage

Source: Dynamic Isolation Systems Inc.

Of all countries, Japan provides the most information about the measured and observed performance of seismically isolated buildings. Its active Japan Society of Seismic Isolation (JSSI) not only promotes seismic isolation but initiates research and publishes reports. In his state-of-the-art summary in 2009, Nagihide Kani (2009) summarized the performance of 23 buildings during six Japanese earthquakes from 2003–08. The recorded movements along the isolation plane ranged from 40–200 mm, with the superstructure accelerations reduced to between 25–50 per cent of the ground accelerations in instrumented buildings.[8]

In 2013, Kani reported this time on the performance of seismically isolated buildings during the 2011 Tōhoku earthquake.[9] Completed questionnaires were obtained from 213 seismically isolated buildings mainly in the most affected Tōhoku and Kanto areas. Of these buildings, 43 are instrumented and 101 experienced displacement along their isolation planes. Typical measured displacements varied from 50–200 mm with a maximum of 415 mm recorded at Miyagi. The maximum residual displacement was around 20 mm. No structural or non-structural damage was reported, apart from two instances of damaged plumbing pipes, but some dampers were damaged. Malfunctioning movement joints were noted in 90 cases. Causes included a lack of maintenance, poor construction and problems with the joints themselves. Remarkably, no building contents or pieces of furniture fell or moved, apart from three items on castors. Building occupants viewed seismic isolation very positively.

Another paper discusses the performance of 20 instrumented buildings in Tokyo ranging in height from two to 21 storeys during the same event.[10] For low values of ground acceleration some buildings behaved, as expected, like conventional structures and amplified the ground accelerations up to a factor of 2.0. But once the peak ground acceleration increased over 0.3 g, the isolation systems reduced superstructure accelerations to approximately 50 per cent of those measured below the isolation plane (Figures 4.6 and 4.7). Another study of the records of eight selected seismically isolated buildings confirms these findings.[11] It notes there was little if any amplification of horizontal accelerations up the height of a seismically isolated building. It can be reasonably assumed then, that unlike a conventional building, the floors of seismically isolated buildings experience similar and low values of acceleration.

This section has summarized the effectiveness of seismic isolation as evidenced from measurements and observations after earthquakes located almost entirely in Japan. The numbers of isolated buildings surveyed, the research findings, including the extremely positive feedback from their occupants, all represent impressive evidence that points to the overall effectiveness of seismic isolation. But before concluding this chapter we return again to the Christchurch Women's Hospital discussed in Chapter 1. Unfortunately it was not instrumented at the time of the 22 February 2011 Christchurch earthquake that caused severe structural and non-structural damage to most buildings in the city's central business district, leading to hundreds being demolished. However the Christchurch Women's Hospital, even though founded on relatively soft soils, was essentially undamaged and functional immediately after the quake. The excellent seismic performance of this building is yet further evidence of the effectiveness of seismic isolation.

4.6 The Yozemi Tower, Tokyo, Japan, which is seismically isolated by elastomeric bearings and oil dampers, displaced 90 mm along its isolation plane. Its ground floor and top floor accelerations were 30 per cent and 45 per cent respectively of the ground accelerations

4.7 Main building, Shimizu Corporation Institute of Technology, Tokyo, Japan, is seismically isolated by six lead-rubber bearings at the top of ground floor columns. The bearings deformed 90 mm and the maximum ground acceleration was reduced by the seismic isolation system to 52 per cent at the second floor and 55 per cent at the top floor

Notes

1 Kelly, T.E., Skinner, R.I. & Robinson, W.H., 2010, *Seismic isolation for designers and structural engineers*, Kanpur: NICEE, p. 154.

2 Sasaki, T., Sato, E., Ryan, K.L., Okazaki, T. *et al.*, 2012, 'NEES/E-Defense base-isolation tests: effectiveness of friction pendulum and lead-rubber bearings systems', Proceedings of the 15th World Conference on Earthquake Engineering, Lisbon, p. 10.

3 Sato, E., Kajiwara, K., Furukawa, S., Ji, X. *et al.*, 2010, 'Full-scaled shaking table test of a hospital made of a base-isolated 4-storey concrete structure', Proceedings of the 9th US and 10th Canadian Conference on Earthquake Engineering, Toronto, paper 303.

4 Moroni, M.O., Sarrazin, M. & Soto, P., 2012, 'Behaviour of instrumented base-isolated structures during the 27 February 2010 Chile earthquake, *Earthquake Spectra*, vol. 28, no. S1, S407–24.

5 Nakamura, Y., Hanzawa, T., Hasebe, M., Okada, K. *et al.*, 2011, 'Seismic isolation and protection, systems: report on the effects of seismic isolation methods from the 2011 Tohoku-Pacific earthquake', *The Journal of the Anti-Seismic Systems International Society (ASSIS)*, vol. 2, no. 1, pp. 57–74.

6 Mayes, R.L., 2007, 'Ronald L. Mayes', *The Structural Design of Tall and Special Buildings*, vol. 16, no. 1, pp. 3–36, p. 18.

7 Seki, M., Yoshida, O. & Katsumata, H., 2012, 'Behaviour of Rikkyo Univ. chapel building retrofitted by seismic isolation in the 3.11.2011 earthquake, Japan', Proceedings of the 15th World Conference on Earthquake Engineering, Lisbon.

8 Kani, N., 2009, 'Current state of seismic isolation design', *Journal of Disaster Research*, vol. 4, no. 3, pp. 175–81.

9 Kani, N., Ogino, N., Kitamura, Y. & Fukazawa, Y., 2013, 'Performance of response-controlled buildings during the huge 2011 earthquake', Proceedings of the 13th World Conference on Seismic Isolation, Energy Dissipation and Active Control of Structures, Sendai.

10 Matsuda, K., Kasai, K., Yamagiwa, H. & Sato, D., 2012, 'Responses of base-isolated buildings in Tokyo during the 2011 great East Japan earthquake', Proceedings of the 15th World Conference on Earthquake Engineering, Lisbon.

11 Iiba, M., Kashima, T., Morita, K., Azuhata, T. *et al.*, 2013, 'Behaviour of seismically isolated buildings based on observed motion records during the 2011 great east Japan earthquake', Proceedings of the 13th World Conference on Seismic Isolation, Energy Dissipation and Active Control of Structures, Sendai.

– 5 –

Benefits and limitations of seismic isolation

Introduction

In the previous chapter we discussed the effectiveness of seismic isolation. Our confidence in the system is based upon three types of evidence: namely computer modelling, physical testing of isolation components as well as full-scale building testing and, finally, measurements and observations made on seismically isolated buildings during damaging earthquakes. Having established that seismic isolation is worthy of our confidence, this chapter explores the extent of its benefits and limitations. Seismically isolated buildings are compared to conventional fixed-base buildings. The following chapters expand upon some of the topics raised here.

The benefits of seismic isolation that make it attractive to building owners are considered in two sections: first, benefits realised before the occurrence of a damaging earthquake, and second, the benefits experienced during and after such an event. As far as limitations are concerned, we explain how for reasons of geology, site location, building height, and maintenance and cost considerations, seismic isolation may not be a sensible strategy, and in some cases, may be impractical to implement.

Pre-earthquake benefits

Surprisingly, seismic isolation can deliver benefits even before an earthquake strikes. Many of these benefits are realized during the design phase of a building when they can lead to improved architectural features. Take, for example, the size of and configuration of structural members. Although seismically isolated buildings are generally designed to be as strong as equivalent conventional buildings, they do not need the same capacity for ductility. Their energy absorption is provided by their damping mechanisms. In conventional buildings ductility is achieved by applying the Capacity Design approach. First, by ensuring a hierarchy of structural damage whereby the most vital structural elements, such as columns that support the whole building weight, have greater protection than say beams. Beams can support their loads even if damaged at their ends. Second, brittle failure modes such as shear failure are supressed by increasing the shear strength of structural members such as columns, beams and structural walls. Then, earthquake energy can be absorbed by the yielding of structural steel, or steel reinforcement in reinforced concrete structures, at non-critical discrete locations throughout the structure.

Such ductile conventional buildings, at least if they rely upon moment frames for their horizontal resistance, require regular inter-storey heights and columns stronger than beams. But since seismically isolated buildings are normally designed to remain elastic, there is not the same need for vertical regularity. Some of the traditional rules for sound seismic configuration can be eased.[1] For example, normally very undesirable soft storey and short column configurations that are normally considered critical structural weaknesses may be possible given that structural members are protected from damage by seismic isolation (Figures 5.1 and 5.2). This potential is exploited in a seismically isolated Japanese building with very slender columns and long-span beams (Figure 3.20).

As well as preventing structural damage, seismic isolation reduces greatly the horizontal inter-storey drifts that occur during an earthquake (Figure 5.3). Where a need to reduce horizontal flexibility rather than meet strength requirements drives the sizes of structural members, seismic isolation enables smaller structural member sizes. Other significant architectural advantages arise in a building subject to less inter-storey drift. These include reduced seismic separation gaps between architectural elements, such as glazing and mullions, or exterior cladding units such as precast concrete panels, and between architectural elements and structural members. The outcome is simplified, less visually prominent construction detailing (Figure 5.4).

Seismic isolation then offers the possibility of more architectural freedom. Structural configurations unthinkable in conventional buildings can be considered. More slender columns and structural walls are also feasible, as well as simpler and thinner separation gaps. These potential benefits can improve architectural aesthetics. Two examples designed by Toyo Ito are discussed in more detail in the following chapter.

But pre-earthquake benefits include more than those related to architectural quality. Seismic isolation offers peace of mind to building owners and occupants. This is reflected in increasing numbers of seismically isolated Japanese apartment buildings. Seismic isolation, marketed as one of the desirable features of these buildings, attracts both tenants and increased rentals. There also may be other financial benefits for owners such as being able to negotiate reduced earthquake insurance premiums or even self-insure, as discussed in Chapter 9.

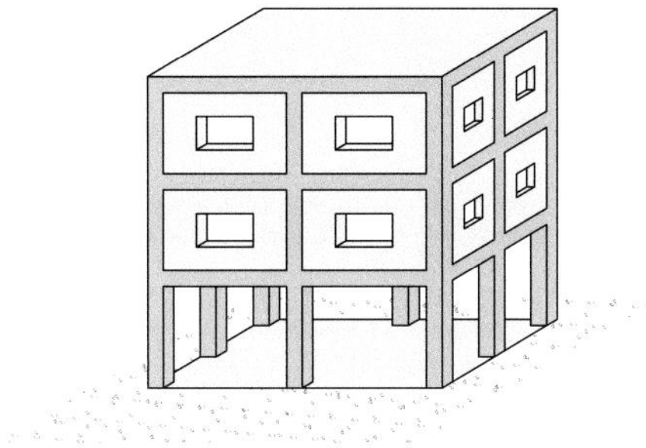

Stiff and strong upper
floors due to masonry infills

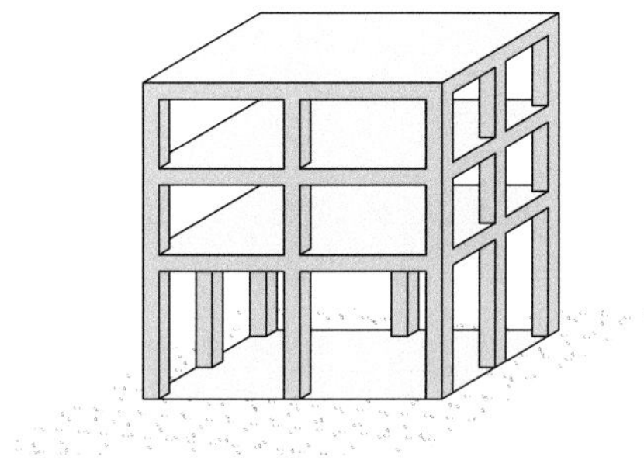

The columns in one storey
longer than those above

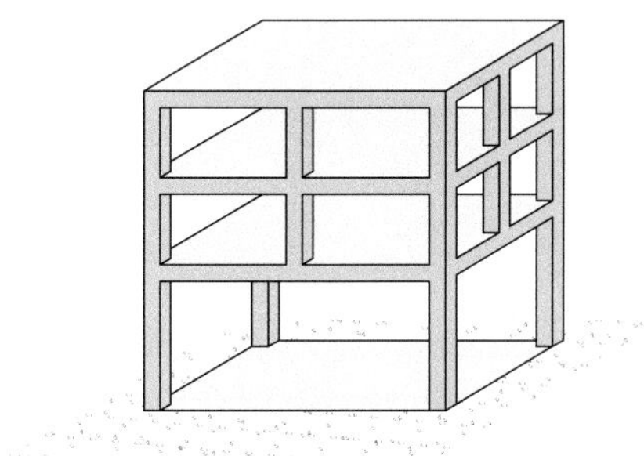

Soft storey caused by
discontinuous columns

5.1 Soft storey structural configurations to be avoided in conventional buildings

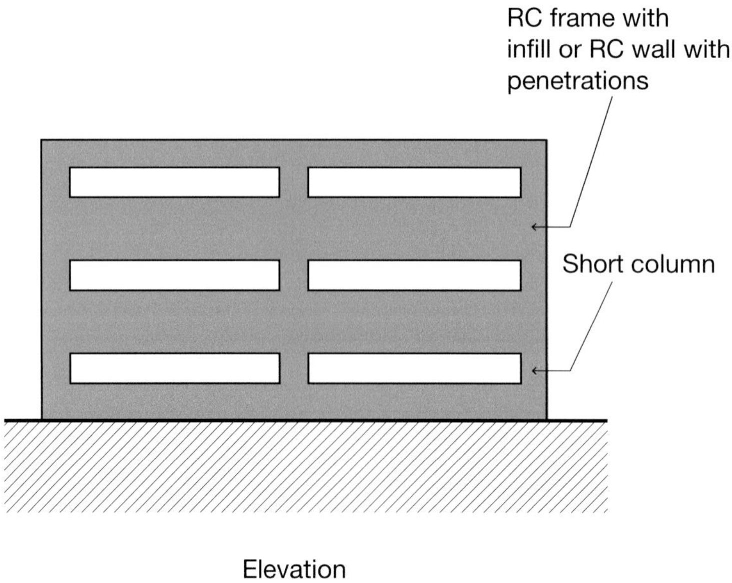

5.2 Short column structural configuration to be avoided in conventional buildings

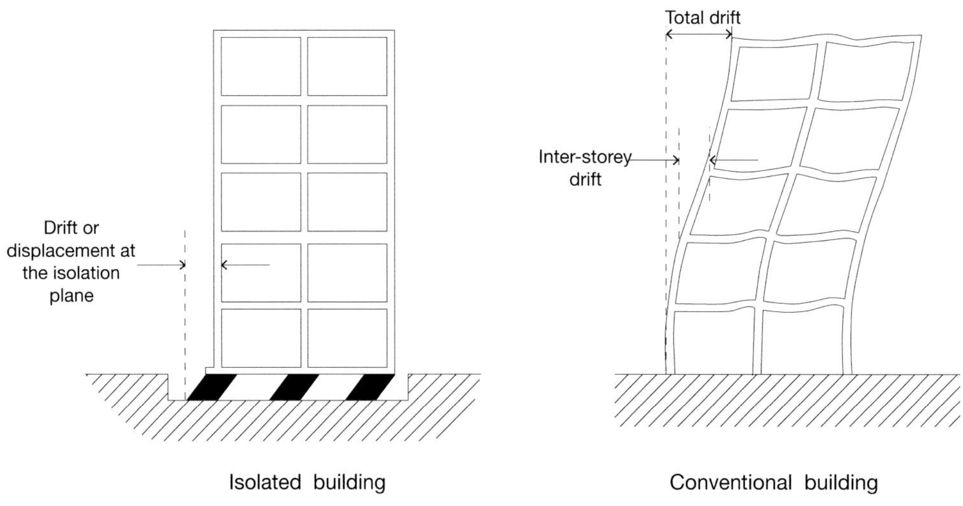

5.3 Inter-storey drifts in conventional and seismically isolated buildings. Most of the total drift of an isolated building occurs at the isolation plane, whereas in a conventional building the total drift is the sum of all inter-storey drifts

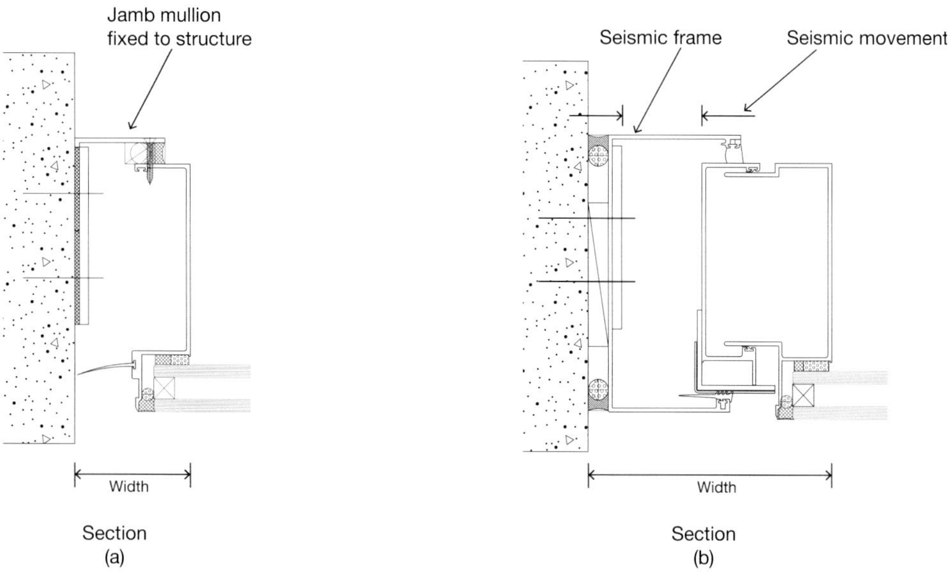

5.4 A horizontal section through a conventional jamb mullion accommodating the small inter-storey drifts of a seismically isolated building (a) compared to a conventional jamb seismic mullion (b)

Benefits during an earthquake and post-earthquake

WHILE the pre-earthquake benefits of seismic isolation are tangible, how much more so are those incurred during and after a large earthquake. They have been referred to in previous chapters especially where comparing the performance of seismically isolated and conventional buildings, but can be explicitly listed as:

- reduced trauma to building occupants,
- reduced injuries to building occupants and passers-by,
- no or minimal structural damage,
- no or minimal damage to architectural (non-structural) elements, and
- no or minimal disruption to building occupancy and function.

It is far less traumatic experiencing an earthquake in a seismically isolated building. The building sways relatively slowly as compared to a conventional building where you are likely to be subject to violent shaking. The lower accelerations in an isolated building also mean less damage to building contents. Less flinging and falling of items from shelves or desk-tops also lowers anxiety levels and reduces injuries. Smaller inter-storey deflections mean glazing is less likely to be damaged or cladding panels are less likely to fall from badly distorted frames onto footpaths below.

Reduced accelerations and deflections are the norm in isolated buildings. For example, Istanbul's Sabiha Gökçen Airport, one of the largest seismically isolated buildings in the world, is predicted to experience up to 80 per cent reduction in inter-storey drifts and floor accelerations (Figures 5.5 to 5.7).[2] In other words, the drifts and accelerations of a conventional building are up to 4.0 times greater. A similar finding emerges in a study of two three-storey

5.5 Istanbul's Sabiha Gökçen International Airport Terminal, Turkey, one of the largest seismically isolated buildings in the world

Source: H. Darama, Arup

5.6 Sabiha Gökçen International Airport Terminal under construction

Source: H. Darama, Arup

steel braced frame buildings, one seismically isolated, subject to many different earthquakes. Its seismic isolation reduced inter-storey drifts and floor accelerations by factors between 5.0 and 20, and 4.0 and 6.0 respectively.[3] The magnitude of these reduction factors vary from building to building and have been found to reduce in more flexible, say moment frame, buildings to a factor of 2.0.[4] The effects of such considerable reductions in drifts and accelerations mean no structural damage and minimal damage to architectural elements and building contents. Scenes like those in Figures 5.8 and 5.9 should never be seen in a seismically isolated building after an earthquake.

5.7 A bearing with its protective covering supports the steel framework

Source: H. Darama, Arup

5.8 Non-structural damage to a suspended ceiling at the Santiago International Airport terminal building after the 27 February 2010 Chile earthquake

Source: G. Mosqueda

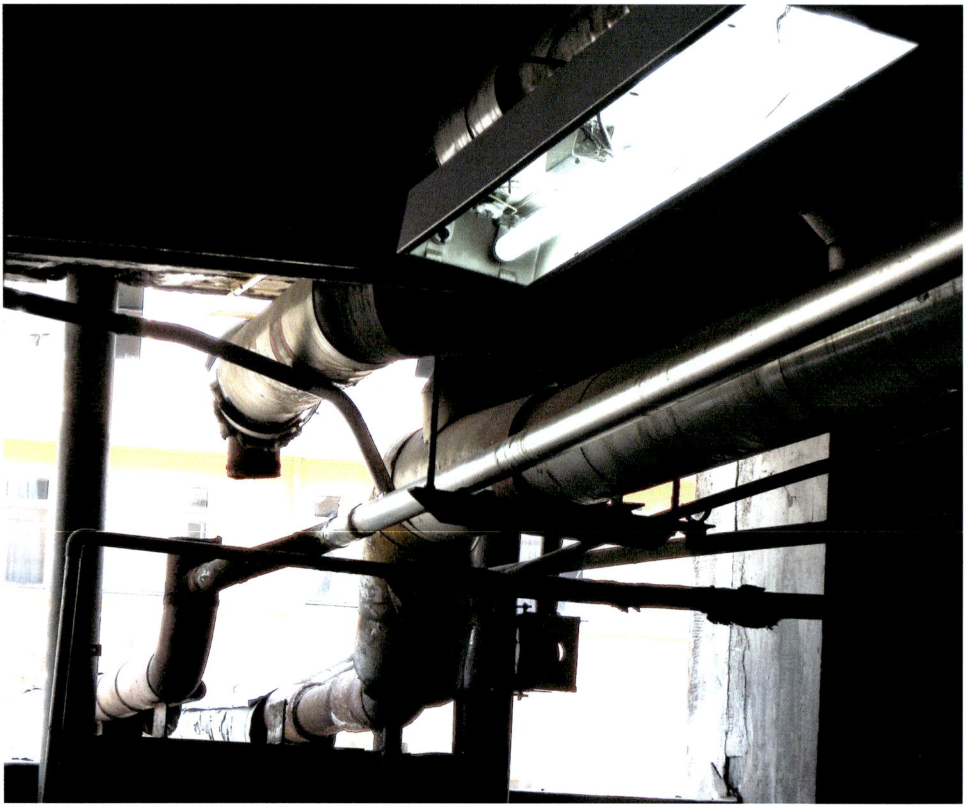

5.9 Damage to ceiling mounted mechanical services at the Santiago International Airport terminal building after the 27 February 2010 Chile earthquake

Source: G. Mosqueda

A consequence of no or limited damage is minimal disruption to building occupancy and function. Of course, the building must remain unaffected by damage to neighbouring buildings and to utilities such as electricity, sewage and water supply, and remain accessible. For some enterprises minimization of downtime is very important. Efforts to quantify post-earthquake costs, including those associated with building downtime, are included in lifecycle cost analyses for seismic isolation systems. These are discussed in Chapter 9.

Limitations of seismic isolation

As we have noted earlier, although seismic isolation is the best strategy to date for reducing earthquake damage in buildings, it does have its limitations. This section begins by explaining how the location and forms of seismically isolated buildings can reduce, or even nullify, the effectiveness of seismic isolation. Then we consider other aspects of seismic isolation that might be of concern to those interested in adopting the technology.

Geological conditions

Like conventional buildings, seismically isolated buildings must be constructed upon adequate ground. Although the vertical forces induced in an isolated building caused by the tendency to overturn under horizontal loading will generally be less severe than for a conventional building, normal rules of geotechnical engineering apply. Not only should sites underlain by fault lines or liquefiable soils be avoided, seismic isolation may not be appropriate for sites with deep soft soils. The reason for this is that such soft-soil sites have their own long natural periods of vibration that can cause seismically isolated buildings to resonate.

Examples of soft soils causing severe earthquake damage to high and flexible buildings (with relatively long natural periods of vibration) are well documented. One of the best known cases occurred in a localized area of Mexico City during the 1985 Mexico earthquake. A small area of the city is built over a former lake bed underlain by soft clay. During this earthquake, whose epicentre was over 350 km away, the soft soil responded to the bedrock shaking like a bowlful of jelly sloshing when shaken. As well as the soft ground itself resonating with a period of approximately 2.0 seconds, flexible buildings between 6 and 16 storeys with a similar natural period amplified that ground shaking even further. Many of these multi-storey buildings were severely damaged or collapsed. Adjacent low-rise masonry buildings that did not resonate remained undamaged. Had a seismically isolated building with a natural period around 2.0 seconds been subject to this shaking, it too would have been severely tested. One of the primary aims of seismic isolation is to lengthen the natural period of the building to avoid those natural periods (or frequencies) of ground shaking that contain significant earthquake energy. Due to the change of shape of a soft soil site response spectrum, shifting the period of the building is counter-productive (Figure 5.10).

Recent research following the 2010/2011 Christchurch earthquakes suggests that seismic isolation with its far higher levels of damping (25 per cent compared to 5 per cent in conventional buildings) is more suitable for soft soil sites typical of Christchurch than previously thought.[5] Clearly, thorough geotechnical investigations and preliminary structural design analyses need to identify any site-specific problems that might indicate that seismic isolation is not as effective as it would be at a firm soil site.

A second and the final geological condition also of concern to designers of seismically isolated structures is the 'near fault' or 'fault-fling' effect. These terms refer to the several high-energy pulses that radiate out from an epicentre during faulting that are especially destructive close to a fault. Unfortunately, these low frequency pulses cause long period structures including seismically isolated structures to resonate—resulting in larger than normal movement along the isolation plane. This might lead to the superstructure colliding with perimeter retaining walls. Buildings less than 2 km from a fault are most affected. Further

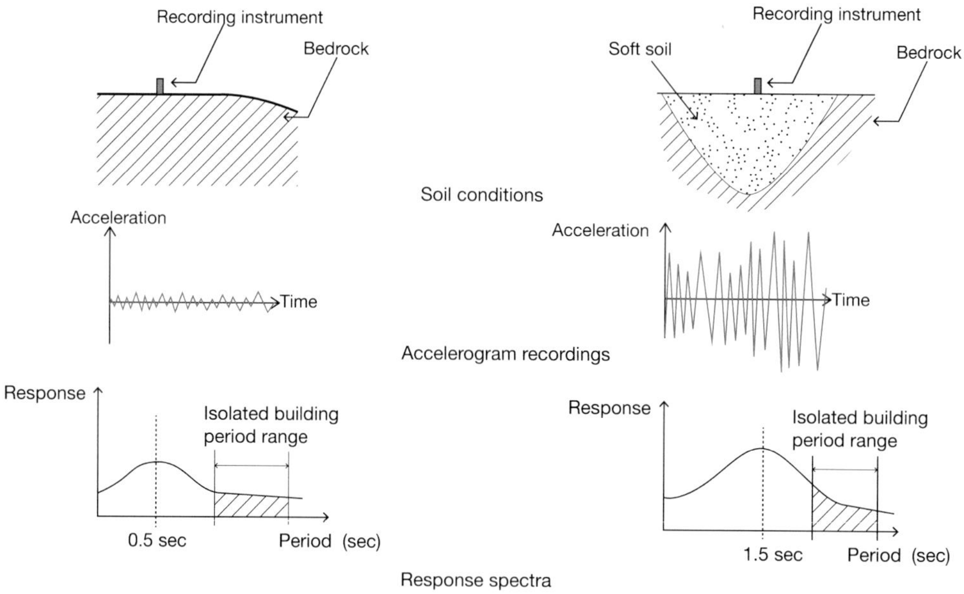

5.10 The effects of soft soil on earthquake shaking. To the left a rock site and to the right, a soft soil site. Compared to rock, soft soil resonates. Shaking intensifies and the natural period at peak response increases to be closer to the range of natural periods of vibration of isolated buildings

away the pulses diminish until they can be neglected at a distance of 20 km. An appreciation of the intensity of this effect can be gained from the fact that conventional buildings in the same period range as seismically isolated buildings, three to five seconds, must be designed up to 70 per cent stronger than low- to medium-rise buildings.[6]

If these near-fault effects could possibly apply to a given design, structural engineers should include several near-fault earthquake records in the suite of records used for design. According to Peng Pan and his co-authors (2005), many Japanese designers had not explicitly designed for near-fault pulses. They believed such an event to be very rare and acknowledged the inevitability of a serious superstructure-perimeter wall collision.[7] The authors then highlight several buildings designed for near-fault pulses where the solution was 'to place a very strong vertical load truss system in the first storey above the isolation level in order to minimize the support points'. Isolators up to 1500 mm diameter sustain the compression stresses and horizontal deformations of more than 800 mm.

As well as experiencing high energy low frequency pulses, buildings located near an epicentre may also likely be subject to significant vertical accelerations. Apart from one 3-D seismically isolated Tokyo building (Figures 5.11 and 5.12), building isolation systems are ineffective vertically. Fortunately, vertical shaking is usually of far less concern than horizontal movements but it may induce some unexpected damage to long cantilever beams and

5.11 Chisuikan, Tokyo, Japan, designed by Kozo Keikaku Engineering and completed in 2011, is a housing complex and the first building in the world to be protected by 3-D seismic isolation

particularly to building contents. The potential damage arising from vertical seismic attack can be considered a limitation of conventional seismic isolation.

Building height

Higher buildings are more flexible against horizontal forces than shorter buildings. As buildings increase in height, so do their natural periods of vibration. If we consider the shape of a response spectrum for firm soils the acceleration response reduces after natural periods exceed approximately 0.5 seconds (Figure 5.13). To a significant degree then, increasing the height of a building tends to isolate it from short period seismic vibrations. Then seismic isolation provides a less dramatic period shift and is less effective as compared to isolating lower-rise buildings.

Whereas the first generation of seismically isolated buildings were low- to medium-rise, they are now increasing in height. There are numerous isolated buildings in Japan over 20 storeys and this trend continues in other countries. For example, a 25 storey office tower

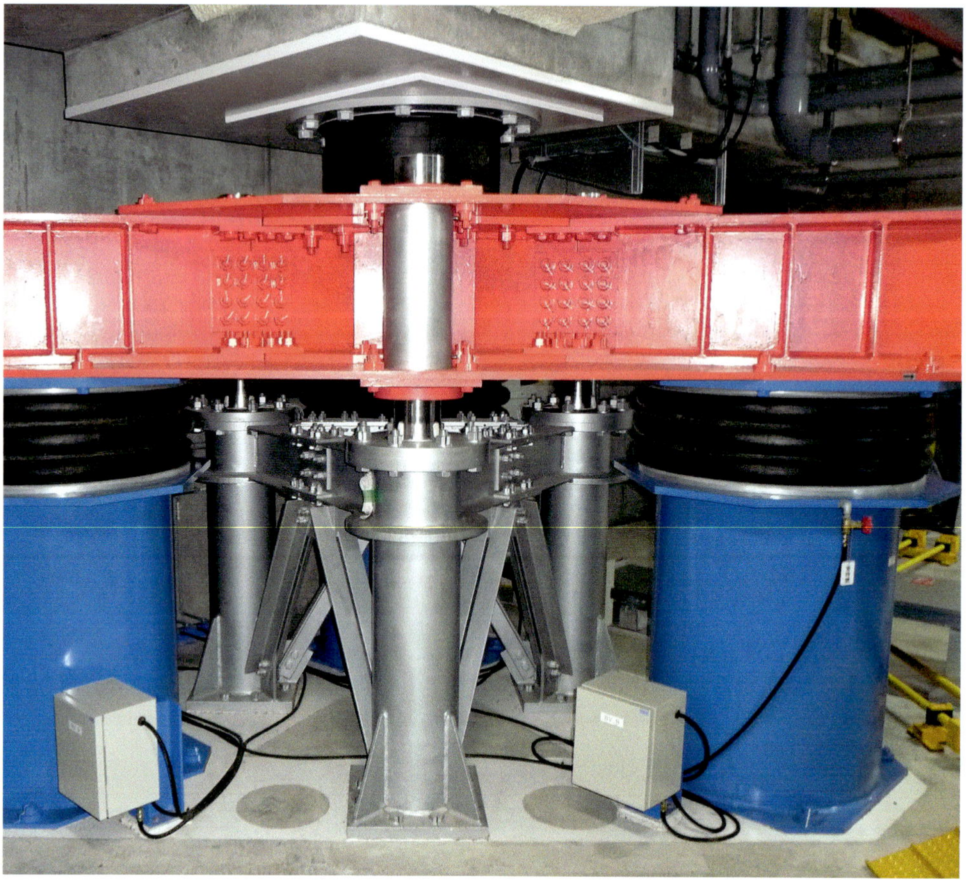

5.12 The 3-D isolation system of Chisuikan. Located in a basement, vertical isolation is achieved through the air-springs atop three blue cylinders and silver-coloured dampers. The bearing above the red beams provides horizontal isolation

has just been completed in Jakarta (Figure 5.14). A fixed-base period of approximately 2.0 seconds was increased by a further 2.5 seconds with isolation. This achieved significant reductions in inter-storey drifts and forces within the building which translate to overall improved seismic performance. Immediate occupancy is expected after the design earthquake.[8] The heights of seismically isolated buildings are not as limited as was once assumed due to advances in bearing manufacture plus designers' choice for the natural periods of isolated buildings to be significantly longer than the 2.5–3.0 seconds of the past.

Just as few very tall buildings are seismically isolated, so too very small buildings, such as houses. Although in principle seismic isolation is applicable to small buildings, and over 4000 seismically isolated houses have been constructed in Japan, a number of factors limit extensive application of the technology to houses. A study of base isolation of timber-framed buildings concludes that because the cost of bracing walls is relatively low, isolation may only

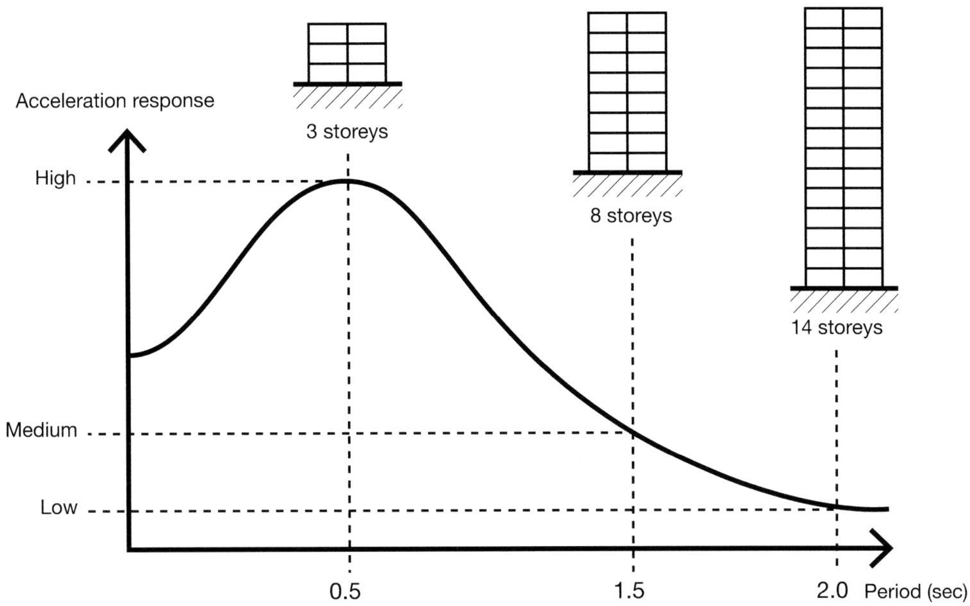

5.13 A response spectrum indicating how the acceleration responses of buildings reduce as they become higher

be justified where there are insufficient bracing walls. Also, unless isolators are placed under a heavy (concrete) suspended floor slab, horizontal deflections in wind storms are excessive.[9] The provision of such a slab increases construction costs significantly. This is why a special synthetic sheet slip layer under concrete foundations is attractive, but a combination of high costs and unconvincing test results is unlikely to see this system applied in practice. Methods to limit wind movement of lightweight houses by 'locking' the isolation system and releasing it only during an earthquake by some fail-safe system were not considered. The study did acknowledge that seismic isolation of light-weight houses might be justified where contents require protection.

A good example justifying the seismic isolation of a small building because of the value of its contents is the modular containerised data centre at Victoria University of Wellington (Figures 5.15 and 5.16).[10] The unit houses the university's backed-up data in the event of its main computer centre being damaged by an earthquake. Allowing up to 500 mm vertical and horizontal movement the QuakeSurfer™ platform provides full 3-D seismic isolation.

Adjacent buildings

The effectiveness and application of seismic isolation can be limited by adjacent buildings. It makes little sense to isolate a building that is close to a more vulnerable building. Damage

5.14 The Gudang Garam Tower, Jakarta, Indonesia. A seismically isolated 25 storey building

Source: S.M. Hussain

5.15 A 3-D seismically isolated back-up data centre, Victoria University of Wellington. The main module is behind the self-contained UPS generator to the left

to that building, by way of falling panels or other elements, could directly damage the isolated building, or could fully or partially destroy the isolation scheme. Debris could fill the horizontal separation gaps, preventing the superstructure remaining isolated. Even in the absence of falling debris, the risk of the damaged building collapsing onto the isolated building might be considered so high as to red placard both buildings until the vulnerable building is stabilized (Figure 5.17).

Nearby buildings can affect other less damaged buildings even if they are further away, such as in the same street or city block. This scenario affected many building owners after the Christchurch earthquakes. Much of the central business district was 'red zoned', meaning public access to even undamaged buildings was prohibited until potential hazards from badly damaged buildings were reduced (Figure 5.18). Although a seismically isolated building in a red zone might survive undamaged, weeks or months might pass before it can be inhabited.

Site coverage

Modern building codes require that buildings be set-back on their sites to avoid pounding neighbouring buildings. Apart from along street frontages, buildings are to be separated far

5.16 A LoGlider bearing supporting the platform. The red mechanical components provide vertical isolation

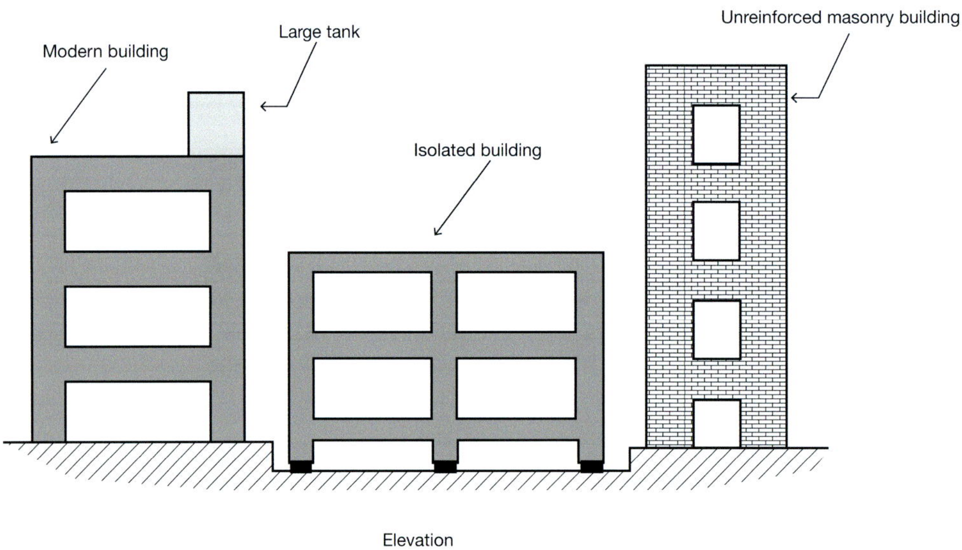

5.17 A seismically isolated building should not be built adjacent to a more vulnerable building. The tank could fall from the modern building to the left, or walls or parapets fall from the unreinforced masonry building to the right, damaging the isolated building

5.18 Buildings in Christchurch cordoned off in the 'red zone' of the city following the 2011 earthquake

enough away from their boundaries so that when they sway sideways they won't damage what may lie beyond (Figure 5.19). The width of the separation gap increases both with the height and flexibility of the building. If a building owner is dismayed at losing so much buildable area on an expensive site, he or she can request a stiffer structure to reduce the width of the gap. For a high-rise building, wide gaps at upper storeys may be easily achieved by designing set-backs. But for seismically isolated buildings almost all of the movement occurs at the plane of isolation. Therefore, irrespective of the building height a wide seismic gap is required at ground level. A 400 mm wide separation gap around three sides of a 10 m by 20 m site reduces the gross floor area by 10 per cent. The need for such a gap rules out seismic-isolation as a retrofitting option in many cases unless a whole cluster of abutting adjacent buildings is tied together to act as a single building and isolated.

Potential cost

The possible increased cost of construction of seismically isolated buildings is a stumbling block for some building owners. Chapter 9 discusses this topic in detail.

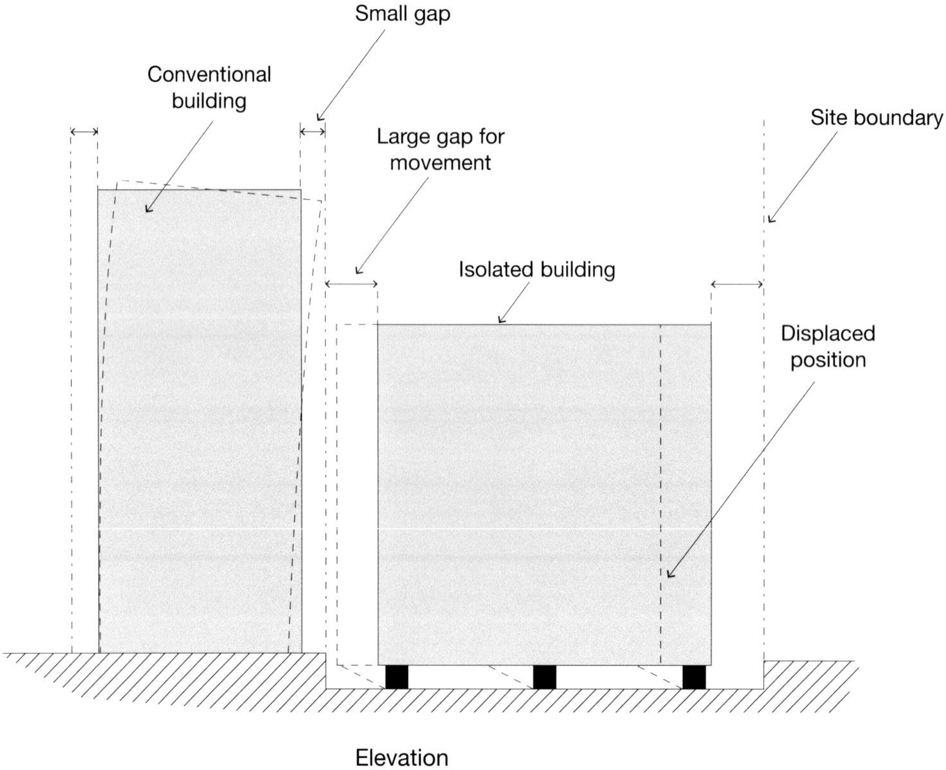

5.19 The need for separation gaps between buildings and boundaries – conventional and seismically isolated buildings

Maintenance

The structural systems of conventional buildings don't require special maintenance. Usually, given the absence of any moving parts or mechanisms, once constructed they can be virtually forgotten. In contrast, as elaborated upon in Chapter 10, isolation devices and design requirements such as movement gaps need rigorous periodic checking. Corroded or dirty slider bearings equate to poor seismic performance, and if seismic gaps are filled by any later construction or storage of materials, the isolation scheme is ineffective. And, if as expected, the structural design has been optimized on the basis of an isolated superstructure, unforeseen severe structural damage is a likely outcome during a large earthquake. Seismic isolation, such as elevators and other mechanical equipment, is therefore suited only for building owners willing to commit to regular inspections and maintenance.

Reduced effectiveness in small earthquakes

Seismic isolation is ineffective during very small earthquakes. During low levels of ground acceleration a seismically isolated building behaves identically to a conventional building. Only when the static friction within a sliding bearing is exceeded, or the lead plug in a lead-rubber bearing yields, does the isolation system 'give' and begin working. Most recently this effect has been measured in the Christchurch Women's Hospital.[11] This means that occupants of seismically isolated buildings feel small sharp earthquake movements just like people in conventional buildings. Once ground motions increase, movements in the isolated building are far reduced in amplitude and are more gentle.

Notes

1 Arnold, C. & Reitherman, R., 1982, *Building configuration and seismic design*, New York: John Wiley.

2 Darama, H. & Zekioglu, A., 2013, 'Implementation of seismic isolation in Turkey for continued functionality', Proceedings of the 13th World Conference on Seismic Isolation, Energy Dissipation and Active Vibration Control of Structures, Sendai.

3 Cutfield M.R., Ma Q.T. & Ryan, K.L., 2014, 'Cost-benefit analysis of base-isolated and conventional buildings: a case study', Proceedings of the 2014 New Zealand Society for Earthquake Engineering Conference, paper O45.

4 Ryan, K.L., Erduran, E., Sayani, P.J. & Dhao, N.D., 2010, 'Comparative seismic response of code designed conventional and base-isolated buildings to scenario events', Proceedings of the 9th US National and 10th Canadian Conference on Earthquake Engineering, Toronto.

5 Whittaker, D. & Jones, L.R., 2013, 'Design spectra for seismic isolation systems in Christchurch, New Zealand', Proceedings of the 2014 New Zealand Society for Earthquake Engineering Conference, Wellington.

6 Standards New Zealand 2004, *Structural design actions NZS 1170:5*, Standards New Zealand, Wellington, 20–1.

7 Pan, P., Zamfirescu, D., Hakashima, M., Nakayasu, N. *et al.*, 2005, 'Base-isolation design practice in Japan: introduction to the post-Kobe approach', *Journal of Earthquake Engineering*, vol. 9, no. 1, pp. 147–71, p. 165.

8 Hussain, S.M., AlHamaydeh, M.H. & Aly, N.E., 2012, 'Jakarta's first seismic-isolated building – a 25 story tower' 'Proceedings of the 15th World Conference on Earthquake Engineering, Lisbon.

9 Thurston, S.J., 2007, 'Base isolation of timber-framed buildings', *Bulletin of the New Zealand Society for Earthquake Engineering*, vol. 40, no. 4, pp. 183–9.

10 [no author supplied] 'Quakesurfer, Victoria University of Wellington.' Available from: www.nzsee.org.nz/wp-content/uploads/2013/05/Quake-Surfer1.pdf (accessed: 23 December 2015).

11 Sridhar, A., Kuang, A., Garven, J., Gutschmidt, S. *et al.*, 2014, 'Christchurch Women's Hospital: analysis of measured earthquake data during the 2011–12 Christchurch earthquakes', *Earthquake Spectra*, vol. 30, no. 1, pp. 383–400.

– 6 –
Seismic isolation and architectural form

Introduction

IN the previous chapter, one of the pre-earthquake benefits noted was the increased architectural freedom seismic isolation offers. The architectural benefits of reduced inter-storey drifts and possible lower levels of structural strength and ductility can be exploited in more elegant detailing, and more slender and less regular structure. This chapter explores these architectural benefits, and more, in greater detail. A series of case studies illustrates how architects are beginning to take advantage of seismic isolation to achieve and reinforce a wide range of architectural objectives.

The case studies are grouped in four categories of architectural concepts and qualities. The categories have been chosen from a total of ten categories that encapsulate the most prevalent architectural concepts and qualities in contemporary architecture.[1] The categories most relevant to the few examples of special seismically isolated architecture are: grounded – floating, stability – instability, heavy – lightweight and simple – complex. Note that each category is defined by two extremes within which a work of architecture can be positioned. Some buildings discussed below reinforce concepts and express architectural qualities from more than one category. Perhaps they could even be placed in another category altogether. Our aim in categorizing buildings like this is not to definitively pigeon-hole them, but to facilitate an exploration of the variety and richness of architectural form aided by seismic isolation.

Grounded – floating

THE case studies of this section lie towards the 'floating' end of the 'grounded – floating' spectrum. This is unsurprising. A building that expresses a sense of being grounded appears to be strongly anchored or fixed to its site. Like a tree rising from the soil, its vertical structure is continuous from the foundations upwards. Any visible horizontal seismic isolation plane breaks that continuity. In many seismically isolated buildings though, the isolation plane lies beneath ground level, out of sight. In this situation it *is* possible to have structure read as grounded, as in the case of the Tod's building.

At Tod's Omotesando Building, Tokyo, the deciduous trees along its street frontage are represented abstractly in the concrete perimeter load-bearing structure (Figures 6.1 and 6.2). The six-storey superstructure rests on elastomeric bearings above a single basement level. Wide 'trunks' at ground level become finer as they rise towards the roof, reducing in width to slender 'branches'. The criss-crossing concrete wall piers form a dramatic surface pattern both inside and out, with exterior surfaces of the structural members flush with the glazing. The structural challenges associated with designing such a unique structure in a highly active seismic zone were lessened by incorporating seismic isolation. Not only could an irregular structural form be realised without making individual members too massive, but the reduction of inter-storey drifts allowed for simple flush-glazed detailing.

Architect Toyo Ito & Associates also exploits seismic isolation in the Tama Art University Library. This building touches the ground lightly rather than appearing grounded (Figures 6.3 to 6.5). The tiny feet of the arches belie the weight of the steel and concrete structure and the library books they support. The library building also avoids the straight lines of Tod's – instead it celebrates the potential of arches to create architecture. Once again, concrete is the dominant material, but at the Tama library, centrally located steel plates within the 200 mm thick arches provide most of the structural strength.[2] In structural terms, the superstructure can be described as two levels of pinned-base steel portal frames, rather than arches. The thickness of the steel plates increases greatly at the bases of the arches whose delicate cross-sections are of solid steel. Another unusual feature of the building is its sloping ground floor in the public areas. To cope with this, as well as a basement over the rear of the site, the isolation plane steps under the ground floor and basement area. Not only does the elastomeric isolation system enable a fineness of structure, particularly at the bases of the arches (or portal frames), but small design-level inter-storey drifts enable flush glazing to combine with surface structure to achieve very smooth planar and gently curved walls.

Having considered one seismically isolated building that appears as grounded as the abstract tree forms on its façades, and another that appears to rest lightly upon the ground, the remaining case-studies in this section unambiguously express how seismic isolation provides architects with opportunities to 'float' buildings above their foundations on isolation bearings.

First we visit the new main building at the Shimizu Institute of Technology campus, Tokyo (Figures 4.7, 6.6 to 6.8). The four-storey superstructure, supported on six massive circular concrete piers, is elevated above a ground floor car park. As well as supporting the entire weight of the building, the piers function as vertical cantilevers in the event of horizontal seismic forces. Substantial foundation structure beneath the piers withstands the tendency to make them overturn when horizontal forces are transferred to them through the lead-rubber bearings mounted at their tops. The corrugated surfaces of the black bearings show the deformation of the layers of rubber between the internal steel plates. Red steel weldments collect forces from the superstructure and transfer them evenly to the upper surfaces of the bearings.

6.1 Tod's Omotesando, Tokyo, Japan, Toyo Ito & Associates

6.2 An area of the front façade

Because the piers are so widely spaced, four-storey deep steel trusses span between them and collect forces from beams supporting the suspended floors. The diagonal members of the perimeter trusses are exposed in places, such as on the front façade of the building. This language of bracing is adopted by secondary or tertiary structure, such as around the main elevators. The elevator structure cantilevers vertically downwards from the isolated superstructure and disappears below ground, separated from the ground by a perimeter moat. An escape stair, its structure concealed by cladding, also cantilevers down from above. Rather than plunging into a moat, the stair stops short of the ground to create the necessary horizontal isolation plane. During the 2011 Tōkohu earthquake a video camera captured the movement of a bearing.[3]

For another example of seismic isolation conveying an impression of 'floating' we visit the main building of the Emergency Management Centre, Foligno, Italy (Figure 6.9). Here seismic isolation is celebrated and elegantly integrated with the overall architecture. The squatness of the three-storey building plus its distinctive dome shape suggests a rigid building with a low centre-of-gravity, well suited to its function as an emergency management centre. The building sits on ten perimeter high-damping rubber bearings. Above each a curved concrete rib rises to connect to a central circular core at roof level. An inner shaft that houses

6.3 Tama Art University Library, Hachioji, Tokyo, Japan, Toyo Ito & Associates. Some of the two main façades

vertical circulation is suspended beneath the first floor slab and, although completely structurally separated from ground level, provides the main access to the building. The seismic isolation system contributes to the experience of entering the building. First, a visitor walks under a perimeter arch and between two bearings. Their significance is heightened by their elevation on concrete plinths and how the curves of the arches above the bearings both direct and force the eye to them. Then, after having passed through the threshold of arches, bearings and plinths, you are surrounded by them, except for the circular entry core at the centre. The sense of 'floatation' is enhanced by the lesser dimensions of the bearings, even though they are protected by rounded covers, as compared to the bases of the double concrete arches. To comply with fire egress requirements an exterior set of fixed-base stairs abut the dome via cable-stayed bridges without compromising the dome's seismic isolation.

An architectural reading of a building 'floating' is usually made only where the visual scale of a building far exceeds that of the building's means of support, such as bearings. For both the previous examples you can stand under the first floor and see the entire superstructure supported by bearings. But in many cases only a glimpse is possible. For example at Te Papa, the National Museum of New Zealand, Wellington, visitors can descend

6.4 The smoothness of the façade looking towards the front entrance

6.5 Interior arches with their fine bases, just inside the main entrance

6.6 Shimizu Institute of Technology, Tokyo, Japan. Some of the piers and isolators supporting the main building

6.7 Structure around the elevators

6.8 External escape stair

6.9 Emergency Management Centre, Foligno, Italy. Overall view including stairs

into a small basement viewing room where several lead-bearings are illuminated, and in other buildings, such as the New Zealand Parliament Building, visitors can take a tour of the basement where many bearings that were inserted during a seismic retrofit can be observed. At the Okumura Memorial Museum, Nara, Japan, a taste of flotation is offered to museum visitors. An elastomeric bearing placed between red painted base-plates is proudly displayed through exterior glazing (Figure 6.10). Seismic isolation technology is treated as one of the museum's precious exhibits.

6.10 Okumura Memorial Museum, Nara, Japan, 2007

Source: Raja Hidzir

Stability – instability

For an example of a seismically isolated building expressing notions of stability we visit the Cathedral of Christ the Light, Oakland, California. Apart from a few tell-tale signs, such as cover plates over seismic separation gaps, the seismic isolation system is well hidden. In fact, its existence is unexpected given the strength and stability expressed by the ground floor concrete walls (Figures 6.11 and 6.12). Exposed both outside and within, they surround the body of the cathedral, providing security. They also support the curved wooden ribs and louvres of the striking fully glazed envelope above. This cathedral replaces one irreparably damaged during the 1989 Loma Prieta earthquake. The structural engineer describes the background to using seismic isolation:

> The project site exists only 4.6 km from the Hayward Fault, one of the most active seismic faults in the world, making the request [for no damage in the Californian Building Code's 'maximum capable' earthquake] substantially more challenging. Ground motions expected at the site could result from an earthquake of magnitude 6.0 or more on the Richter Scale. The approach to design embodies lightness (both visually and structurally) and luminosity, hence its name, the Cathedral of Christ the Light. The structure is defined by sacred geometries and designed for lightness by using seismic isolation to manage ground motions and reduce the demand on the superstructure during an earthquake.[4]

For a seismically isolated building that reinforces architectural qualities at the other end of the stability – instability spectrum we return to the Sendai Mediatheque, Japan (Figures 3.15 and 3.16). Chapter 3 explains how ductile steel moment frames in the basement constitute the bespoke isolation system. By absorbing much of the earthquake energy in the moment frames it is possible for the exposed fine and slender structure above ground to avoid seismic damage while displaying instability (Figure 6.13). In interior spaces, sloping column clusters support gravity loads. Four larger and braced cores located near the corners of the buildings not only brace the unstable sloping clusters but resist the seismic forces within the superstructure.

Heavy – lightweight

Like the Cathedral of Christ the Light, the Cathedral of Our Lady of the Angels, Los Angeles also replaced its predecessor, badly damaged in the 1994 Northridge earthquake (Figure 6.14). At the outset of the design the client requested that the cathedral 'should survive even a big quake, unscathed'.[5] This was a challenge given the proximity of an active fault to the site. The architectural response was to resort to concrete walls and seismic isolation. Structural walls dominate the architecture and give a sense of heaviness and solidity. As for most

6.11 The Cathedral of Christ the Light, Oakland, California, USA, Skidmore Owings & Merrill, 2008. Enclosing concrete walls express security

6.12 View of interior and deep concrete walls

seismically isolated buildings, the isolation system is totally concealed. The high walls are tied together at roof level by a thick reinforced concrete diaphragm and the whole complex is supported by almost 200 bearings. Most are high-damping rubber bearings, but slider bearings are also used. Friction pendulum bearings support the four corners of the bell tower and have been designed to withstand vertical tension forces resulting from seismic overturning moments.

While there are indeed some examples of seismically isolated 'heavy' architecture as we have just seen, it is more common to encounter isolated buildings taking advantage of the lighter and more slender structure that seismic isolation can facilitate. The following four case-studies illustrate this trend.

The design of Sony City, Tokyo, was led by both the architect's concept for an expressed perimeter diagrid structure integrated with a double skin façade, and the client's request for 'exceptionally high resilience in strong earthquakes'.[6] Both of these requirements were realized with seismic isolation. Given the inherent stiffness of diagrids against horizontal seismic forces, without an isolation system the member sizes would have been 'heavy and un-economical'. After all, the building is essentially a large cube, with plan dimensions 100 m by 70 m and earthquake forces are resisted only on the four sides of the building (Figures 6.15 and 6.16). The majority of the gravity loads are carried by internal steel frames. Seismic

6.13 Sendai Mediatheque, Sendai, Japan, Toyo Ito & Associates. An 'unstable' column cluster

6.14 Cathedral of Our Lady of Angels, Los Angeles, USA. Rafael Moneo, architect, 2002. Some of the heavy interior elements

isolation is provided by 200 elastomeric bearings, the majority of which are high-damping bearings, and 40 viscous dampers located around the perimeter where they are most effective in dampening torsional vibrations. Each of the heavily laden interior columns is supported by four bearings.

Although far smaller than Sony City, the Prada Boutique Aoyama, Tokyo, is another example of a perimeter diagrid or lattice structure integrated with the façade also resisting most of the seismic forces (Figures 6.17 and 6.18). It is hard to imagine load-bearing walls with a greater degree of transparency, yet still able to withstand both vertical and horizontal forces. Not only does structure allow interior spaces to be flooded with light, the diamond-shaped structural form synthesises with the five-sided crystalline architectural form. One reviewer comments that the architects have noted that this was their first building 'to forge structure, space and facade as a single unit'.[7]

According to the structural engineer, the design concept and structural requirements presented quite a challenge:

1 The diagonal lattice forming the outer lattice should form the structural framework while being an integral part of the glass façade (avoiding the need for vertical and horizontal members in the vertical planes).

6.15 Sony City, Tokyo, Japan, 2006. Plantec Architects. The perimeter diagrid bracing

6.16 At the base of the building the diagrid continues just at the corners, but additional internal bracing is provided

6.17 Prada Boutique Aoyama, Tokyo, Japan, Herzog & de Meuron, 2003. The main entry and the lower floors of the six storeys above ground level. The black glazing lines delineate the diagonal structural members while the floor structure is de-emphasised by a light-colour

2 The diagonal lattice members should be 250 mm in width and 300 mm in depth, including the dimensions of the finish materials.
3 The interior vertical shaft should be made as small as possible (without large horizontal rigidity), and atriums should be provided by floor openings on the 1st, 2nd, 4th and 5th floors.[8]

Of particular concern was the need to brace the building using a system that was not only stiff and therefore needed to be stronger than usual, but also lacked ductility, leading to even higher design forces. Seismic isolation was the only way to meet the dimensional limitations of the architects and to avoid chunky bracing members which would have compromised the architectural concept. The diagonal members are welded I-sections with varying web and flange thicknesses. They are welded to cast steel joints which are also connected to perimeter floor beams. The isolation plane is beneath the basement floor level. It contains 39 bearings – a combination of lead-rubber bearings, rubber bearings and sliding bearings (Figure 6.19).

6.18 The structural diagrid wall is most clearly visible from within the building

As well as facilitating non-conventional structural solutions, seismic isolation is often used in conjunction with moment frames. But in the Nicolas G. Hayek Center, Tokyo, little of the architecture including its moment frames is conventional (Figures 6.20 to 6.22). The frames can be described as 'mega' or 'super' frames, each 'storey' of which contains three or four normal height storeys. The mega-frame structure arose out of the architectural concept for vertical (sky) gardens extending the heights of four atria, one in each 'storey'. The large inter-storey heights of the mega-frame and a lack of internal columns mean that it is extremely flexible during earthquake shaking. Closely spaced (2.4 m centres) columns along each side of the building that form the mega-frames help stiffen it up, but that was not enough. Some form of seismic isolation was required. The structural engineer comments that: 'To achieve stringent performance targets, base isolation would be the usual choice, but as Ginza real estate prices are among the highest in the world, allocating a 1.0 m strip of clearance around the perimeter to absorb the base-isolation movement was deemed inappropriate.'[9]

The final solution involved seismically isolating about half the floor area of four upper level slabs by separating them from the structure around them and placing them on high-damping and slider bearings. As Naomi Pollock explains: 'Perched on rubber bearings, these upper slabs are completely detached from the main frame, enabling them to dampen building

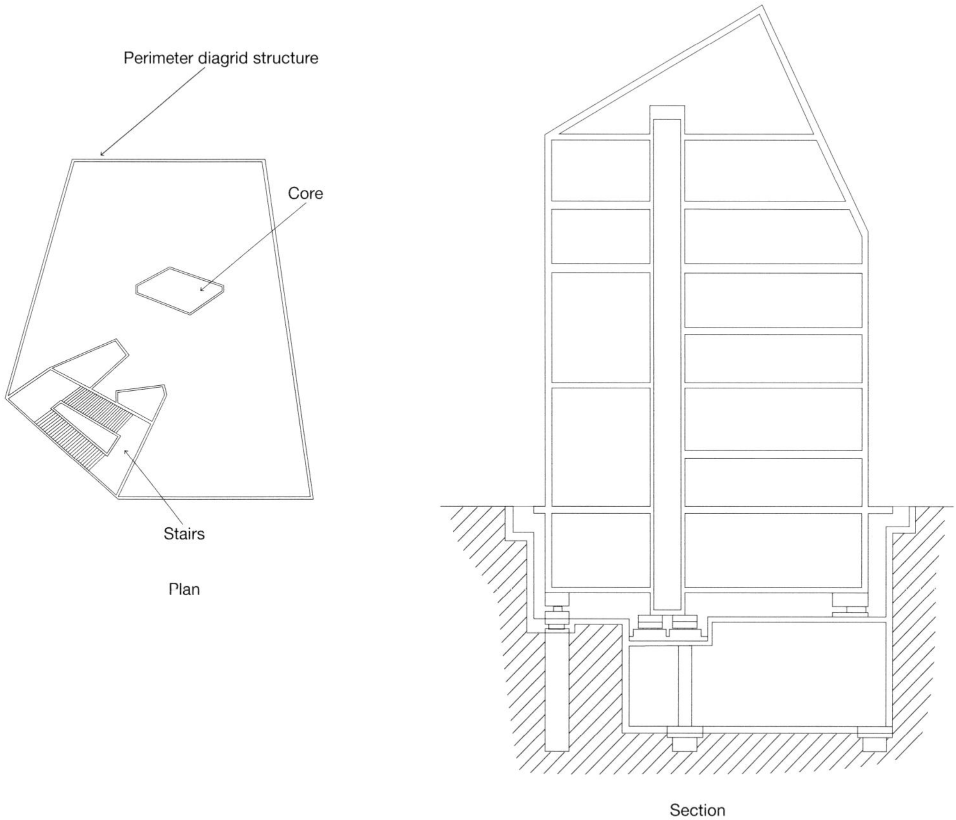

6.19 Simplified floor plan and section

movement and counteract seismic forces by sliding back and forth like a pendulum.'[10] This special type of 'mid-level' isolation proved effective in decreasing the seismic response of the building for all the seven earthquake records used in the design of the building. Horizontal displacements were not a huge issue due to the city council minimum separation between building and boundary in the order of 500 mm. Without seismic isolation the dimensions of the structural frames would have had to be significantly larger.

Light-weight is an apt description of the main space of the seismically isolated San Francisco Airport international terminal. The terminal, underlain by approximately 40 m of soft soils, is built on concrete piers above an existing road (Figure 6.23).[11] Friction pendulum bearings are placed on top of the piers and below the isolated superstructure. Above the isolation plane, two floors accommodate arrivals and baggage handling. These lower floors are braced by conventional structural systems that are protected from damage by the isolators. Departures occur from the 'Great Hall'. Its high enclosing glazed walls and 3-D roof trusses that are fabricated from relatively small cross-sectional steel members create an atmosphere

6.20 Nicolas G Hayek Center, Tokyo, Japan. Shigeru Ban Architects. 2006. Front façade

6.21 The ground floor atrium rises three storeys

6.22 Another atrium near the mid-height of the building

6.23 San Francisco Airport international terminal, USA. A section of the main elevation

of lightness (Figures 6.24 and 6.25). The five lines of trusses are supported by 20 steel cantilever columns whose slenderness reflects the lack of roof-top seismic force amplification as a result of the seismic isolation. The whole structure, including the steel box columns that are clad to give the appearance of circular columns, was designed to a very high level of seismic performance. All of the upmost structural elements will remain elastic (no structural damage) during an earthquake with a 1000 year return period. Seismic isolation achieved the lowest construction cost while meeting the required seismic performance.

Simple – complex

LIKE most conventional buildings, the majority of seismically isolated buildings possess a regularity and simplicity of form. There is little point in illustrating such common architectural forms. So the following case-studies, whichrange between simple and complex forms, begin with a rectilinear form with a difference. The high narrow building has a penetrated and stepped façade, injecting variety into its form. The Yoyogi Seminar School building, also known as the Yozemi Tower, Tokyo, is 26 storeys high above three basement levels (Figures 4.6, 6.26 and 6.27). Pairs of structural walls coupled by diagonal braces that

6.24 A view along the front of the building showing the cantilever columns

6.25 The three-dimensional steel roof trusses that flood the Departures Hall with light

resist transverse horizontal forces are exposed at each end of the building. Interior braced frames act in the longitudinal direction. The isolation system consists of bearings and dampers. Columns bear on a combination of rubber and sliding bearings while damping is provided by 12 semi-active and 12 passive hydraulic dampers. The semi-active dampers adjust the level of damping during an earthquake to optimize seismic performance.[12]

The Inagi Hospital, Tokyo, has a far more complex plan. The irregular tower block form is butterfly-shaped (Figures 6.28 to 6.30). Podiums at the front and the rear of the two wings result in a rectangular ground floor plan. The whole plan area is seismically isolated on bearings below the basement level. If this building were of conventional construction it is likely that due to its re-entrant corners the tower would have been separated into three structurally separate buildings – the two wings and the central core. Especially given the large penetrations for vertical circulation through the central core floor diaphragms, a strategy of separation would have made far more sense than trying to tie the three blocks together through the weakened central area. That is the usual structural response to re-entrant corners.[13] However, since the whole hospital is seismically isolated, its inter-storey drifts and accelerations have been reduced sufficiently for the designers to tie the superstructure together. Rather than vibrating as three independent structures, it acts as a single unit. One of the big advantages of this approach has been to avoid seismic separation gaps between the three towers and the subsequent considerable simplification of the architectural and services detailing across the gaps. For example, the need for floor, wall and ceiling cover plates are avoided, and services do not

6.26 Yoyogi Seminar School building, also known as the Yozemi Tower, Tokyo, Japan. Taisei Corporation, 2008

6.27 A section of the cafeteria

6.28 Inagi Hospital, Tokyo, Japan, front entrance

6.29 Rear elevation

require any flexible connections to accommodate the relative horizontal movements between blocks. Also, the connections between the podiums and the towers have been made a lot easier. Without isolation, the podiums would have had to be tied to one block and seismically separated from the others. Alternatively, they could have been designed as free-standing structures with their own seismic resistance and separated from all three towers. The elegance of the seismically isolated solution is highlighted when these alternative design approaches are considered.

In the final example, seismic isolation is the means of achieving architectural complexity. This quality of the Delegation of the European Union to Japan, Tokyo, is apparent on the main façade notably its randomly placed and sized openings. The seismic isolation scheme comprises natural rubber laminated bearings and sliding bearings. Six oil dampers acting in each orthogonal direction provide restraint against wind, provide damping and control torsional deformation during an earthquake that arises from the 3-D structural complexity.[14] According to T. Mizutani (personal communication, 15 August 2014), due to seismic isolation, which greatly reduced the seismic design forces to approximately 25 per cent of what would normally be designed for, it was possible to design and detail the very complex perimeter and internal reinforced concrete walls satisfactorily. Nonetheless, the highly penetrated walls and other structural complexities such as some walls being off-set in plan, required sophisticated computer modelling. The wall reinforcement was able to be detailed bearing in mind the need for construction practicality.

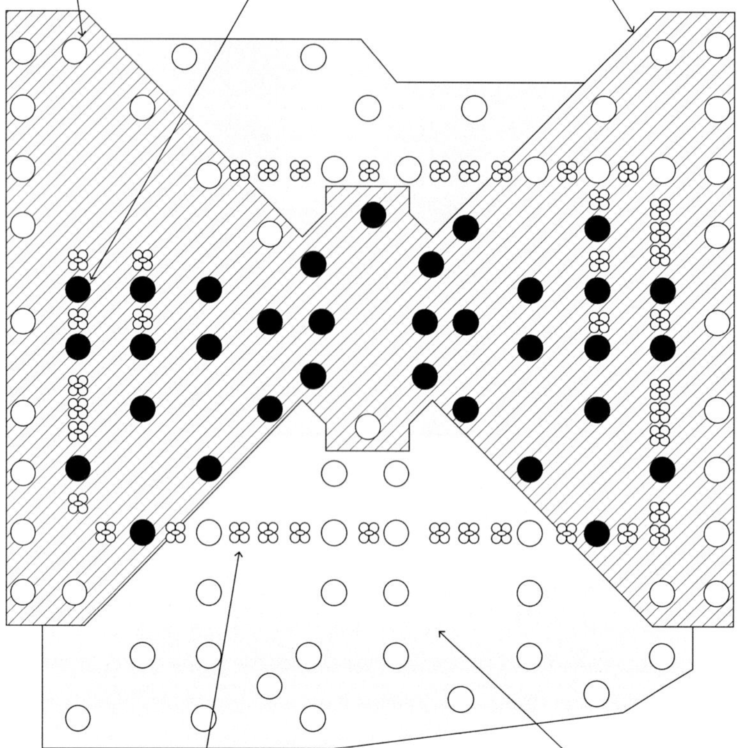

6.30 Plan of the hospital

Notes

1 Charleson, A.W., 2014. *Structure as architecture: a source book for architects and structural engineers*, 2nd edn, Oxford: Routledge, Chapters 12 and 13.

2 Toyo Ito & Associates, 2007, 'Tama Art University Library'. *Japan Architect*, no. 67, pp. 120–29.

3 Shimizu Corporation, 'Effects of seismic isolation.' Available from: www.shimz.co.jp/english/theme/earthquake/effect.html (accessed: 14 August 2014).

4 Sarkisian, M., Lee, P. & Long, E., 2011, 'A celebration of structure as architecture – The Cathedral of Christ the Light', Proceedings of the ASCE, Structures Congress, Buildings section, pp. 971–82, p. 971.

5 Rosta, P. & Post, N., 2001, 'L.A.'s third millennium Noah's ark', *Engineering News Record*, vol. 42, no. 5, p. 42.

6 'Sony City, Tokyo: a diagrid combined with base isolation', *The Arup Journal*, 2009, no. 2, pp. 49–51, p. 51.

7 Herzog & de Meuron 2004, 'Prada Aoyama Epicenter', *A+U: Architecture and Urbanism*, vol. 406, no. 7, pp. 78–83.

8 Nakai, M., 2008, 'Unique architectural forms enabled by base-isolation', Proceedings of the 14th World Conference on Earthquake Engineering, Beijing.

9 'Nicolas G. Hayek Center, Tokyo', *The Arup Journal*, 2009, no. 2, pp. 52–4, p. 52.

10 Pollock, N.R., 2008, 'Shigeru Ban sets showrooms and facades in motion at the Swatch Group's Nicolas G. Hayek Center in Tokyo', *Architecture Record*, no. 5, pp. 200–05, p. 202.

11 Mokha, A.S. & Lee, P., 1999, 'Wings of isolation: San Francisco International Airport's new terminal is protected by 267 steel seismic isolators', *Modern Steel Construction*, October.

12 EERI reconnaissance team 2012, 'Performance of Engineered Structures in the M_w 9.0 Tohoku, Japan, Earthquake of March 11, 2011', *EERI special earthquake report*, pp. 1–16, p. 12.

13 Charleson, A.W., 2008, *Seismic design for architects: outwitting the quake*, Oxford: Elsevier, pp. 132–3.

14 Mizutani, T., Hayabe, Y. & Yoshikawa, H., 2013, 'Large structural wall system with random openings realized by seismic isolation system', Proceedings of the 13th World Conference on Seismic Isolation, Energy Dissipation and Active Vibration Control of Structures, Sendai.

– 7 –

Retrofitting

Introduction

As mentioned in Chapter 2, seismic isolation is a relatively new technology. The first building to incorporate all necessary engineering characteristics was completed in 1981. That building, the William Clayton Building in Wellington, New Zealand, was a new building. The world's first seismically retrofitted building was completed another eight years later. We might have expected the introduction of this new technology to an existing building to be undertaken on a small, simple and modest building. But the Salt Lake City and County Building, USA, is just the opposite (Figure 7.1). It is a large, imposing and historic building. Completed in 1894, and constructed from unreinforced masonry and sandstone, it suffered minor damage during a 1943 earthquake when several statues fell. In the mid-1970s engineering attention focussed on strengthening the vulnerable clock tower with internal steel braces, but later reviews recommended seismic isolation, which was eventually adopted.[1]

This monumental and ornamented building is of masonry load-bearing wall construction. Exterior and interior wall thicknesses up are to 0.9 m and 0.6 m respectively, while at the base of the tower, walls are a massive 2.5 m thick. Three different retrofitting designs were developed, including one which would have entailed interior demolition and reconstruction. Seismic isolation proved the least architecturally disruptive approach while minimizing damage during the design earthquake. Even so, the clock tower required new steel braced frames to resist the seismic forces already greatly reduced by seismic isolation. Other strengthening of the superstructure involved tying back ornamental features, tying walls to floor diaphragms and providing new roof diaphragms.

The isolation plane is just above the original foundation footings on which over 400 lead-rubber bearings were placed. Weight transfer from the masonry walls to the bearings is achieved by two beams, one on each side of the wall and clamped against the walls by tensioned tie-bolts. Short cross-beams transfer loads from the pairs of beams to the bearings centred between them. A new concrete ground floor slab diaphragm ties the whole building together just above the isolation plane, and a continuous horizontal gap between walls and foundations ensures that the bearings are the only contact the building has with the ground.

7.1 Salt Lake City and County Building, Utah, USA. The first building to be retrofitted using seismic isolation

Source: Daderot

Since this first seismic isolation retrofit project, hundreds of others have been undertaken in seismically prone areas around the world. Some projects have been even more impressive in terms of architectural monumentality and scale than the Salt Lake City and County building. Others have little architectural merit and are retrofitted for other reasons that are listed in Chapter 2. The two most significant advantages of this retrofitting technology are first, being able to use the building during the retrofit process, and second, keeping the superstructure retrofit interventions to a minimum.

It is often inconvenient and costly to vacate a building during a retrofit but, with the choice of seismic isolation, many buildings can be occupied during construction. In Japan, approximately 70 per cent of seismically isolated retrofitted buildings continue to be occupied during the retrofit process.[2] Heritage buildings, in particular, benefit from seismic isolation due to the need for minimal strengthening work to their superstructures. For example, at the Salt Lake City and County building, major new structural components were needed only in the clock tower.

The next section discusses options to be considered when deciding upon the height or level of the isolation plane in a retrofitted building. Then we present several case-studies of completed projects, dividing them into two groups – unreinforced masonry buildings, and then more recent column and beam framed buildings. The chapter concludes by discussing limitations associated with seismic isolation retrofitting.

Location of the isolation plane

THE height of the isolation plane above the top of existing foundations depends on a number of factors, such as the type of construction, and whether or not there is a basement. Isolation plane heights tend to be lower in unreinforced masonry buildings compared to frame buildings. Masuzawa and Hisada report that in the 90 isolated retrofitted buildings they studied, 40 per cent had the isolation plane near the existing foundations, and in the remainder the isolators were placed in between floors.[3] Where located between floors, 40 per cent were in basements, 40 per cent in ground floors, and the remainder at higher levels.

In existing load-bearing wall buildings, mainly constructed from unreinforced masonry, isolation bearings are placed under and within walls. Slots are cut through the walls where bearings need to be placed. The use of discrete bearings raises two structural challenges. First, new beams must pick up the vertical loads from the walls above the bearings and transfer them to the bearings. In most cases 'sandwich' beams, cast on both sides of the walls and then clamped together with tensioned rods, as at Salt Lake City, perform this role. If, for aesthetic or heritage reasons, this solution is inappropriate, it is possible to cast a new beam, section by section, directly under an existing wall but this procedure requires extensive propping of the wall and is to be avoided if possible. Just as beams transfer forces from above into the bearings, similar beams are usually required under the bearings to safely distribute the point loads from individual bearings to the lengths of foundation walls underneath them.

A basement makes the isolation of a load-bearing wall building much more feasible. As shown in Figure 7.2, if a building is founded on shallow foundations it might be necessary to raise the level of the ground floor to accommodate the depth of the upper sandwich beams (approximately 1.0 m) and the height of the bearings (between 300–600 mm) as well as create a crawl space for construction and maintenance access. If the foundations are deeper, the bearings and the two layers of sandwich beams can be constructed below the existing ground floor but this requires a large volume of excavation. With a basement, the construction of the beams, especially alongside interior walls, is so much easier. The upper sandwich beams are normally cast near the top of the basement walls. This minimizes bending of the upper walls when the bearings are displaced normal to the walls. New concrete wall jackets may be required either side of the basement walls to withstand far greater bending from the same horizontal forces acting normal to their lengths. Existing internal masonry or concrete columns in the basement will have bearings placed above them and they will also require

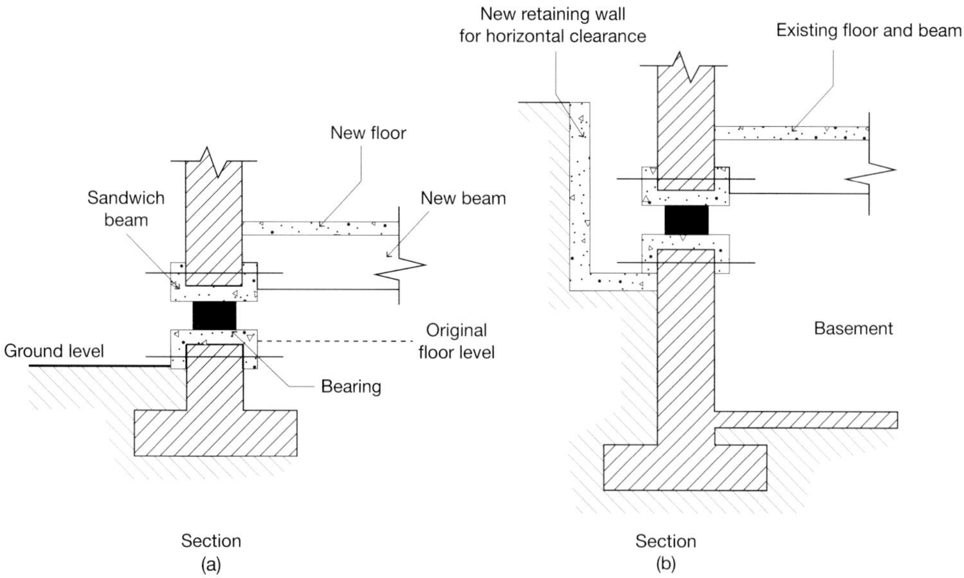

7.2 The advantage of a basement in isolating an existing building is due to the extra height in which to create the isolation plane. In (a) construction conditions are very cramped, while in (b) the basement makes construction easier

jacketing and possible enlargement of their footings to resist the bending at their bases. Figures 7.3 and 7.4 show typical construction details.

Retrofitting concrete or steel frame buildings by seismic isolation is easier than for load-bearing wall buildings. There is far less structure to isolate and there is usually no need for beams to transfer loads from masonry into and out of bearings. First, a decision must be made concerning the level of the isolation plane. We need to bear in mind the horizontal plus or minus 300 to 500 mm design movement in the isolation plane and its implication for elements, such as elevators, stairs, non-structural walls and services, crossing it. More information about treating these elements is given in Chapter 8. If we assume the isolation plane will be somewhere within the ground floor level there are three options (Figure 7.5). Insertion of bearings at column bases is less common due to the high degree of bending in the column (and beam) at first floor level. In many situations this bearing placement option requires significant structural strengthening of the first floor structure. Mid-height isolation is more common. The existing column, beam and foundation strengths may be sufficiently strong. If not, bearings might be best placed just below the first floor beams. Now the bending demands on the first floor beams and columns are as low as possible. However, the ground floor columns and their foundations may need upgrading to withstand the greater bending at foundation level. The construction sequence for isolating an existing frame building is similar to that shown in Figure 7.4.

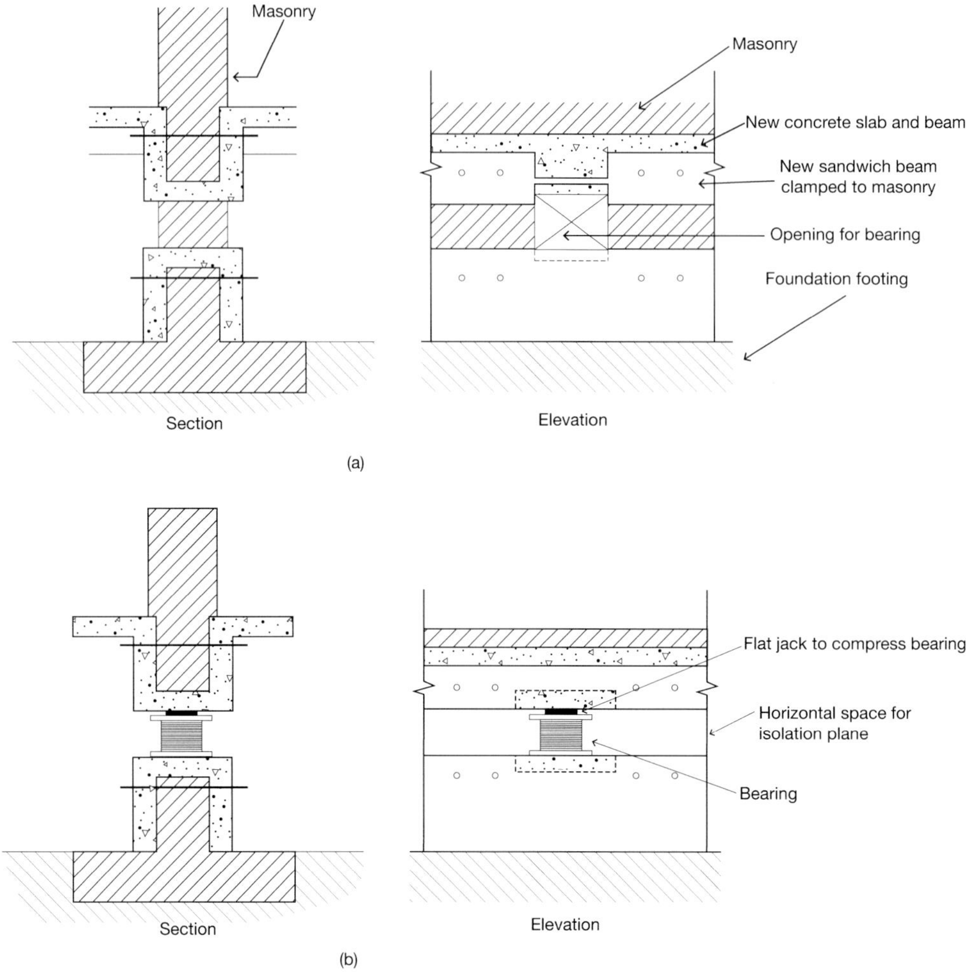

7.3 Typical construction details when inserting isolation bearings in a load-bearing wall without a basement. (a) New sandwich beams are constructed and an opening for the bearing is made, and in (b) the bearing is inserted, pre-loaded by a flat jack to carry the loads from the wall

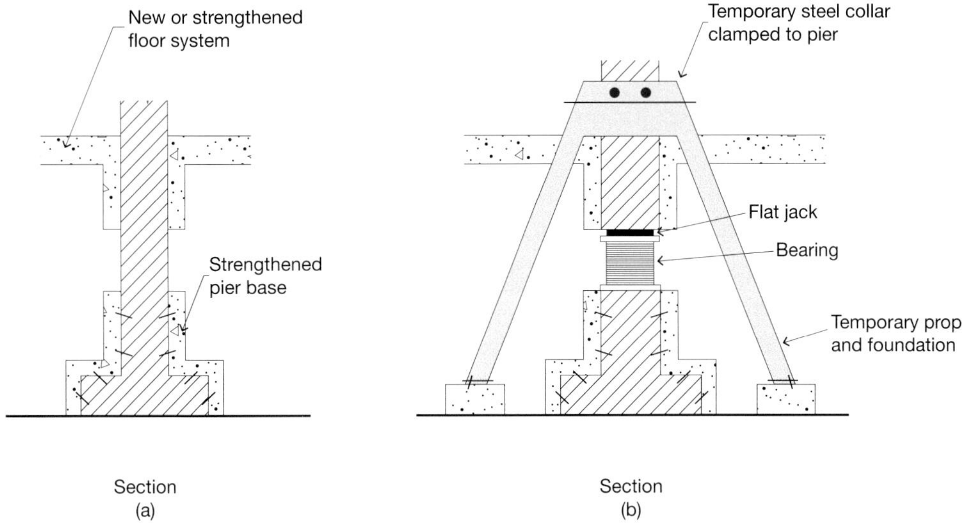

7.4 Typical construction details when inserting isolation bearings under a masonry column or pier. After strengthening the pier base and forming a strong slab-and-beam floor (a), the pier load is transferred through temporary props so the pier can be cut and a bearing inserted and then loaded with a flat jack (b)

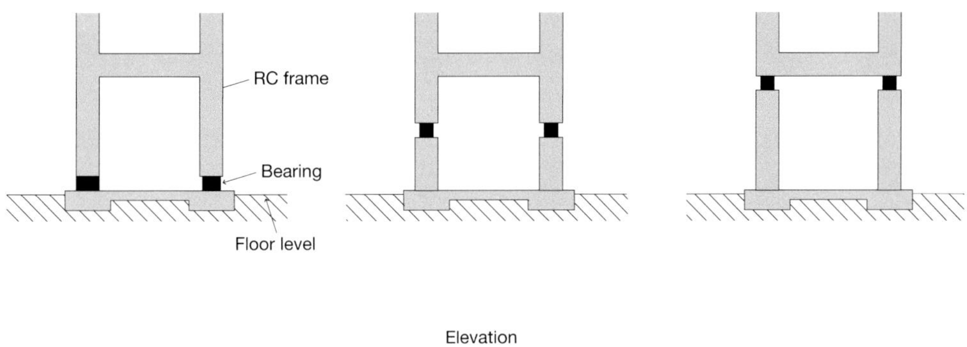

7.5 Three potential levels of bearings and isolation planes for an isolated retrofitted moment frame building

Case-studies of retrofitted unreinforced masonry buildings

International Library of Children's Literature, Tokyo

This was the first unreinforced masonry building in Japan to be retrofitted using seismic isolation (Figure 7.6). The main body of the building, constructed from load-bearing brick walls and interior steel posts and beams, was completed in 1906. In 1929 it was extended using reinforced concrete frames with brick infill.[4] 2002 marked the completion of the most recent extension, refurbishment and retrofit. The renovation of this designated historic building required upgrading of its seismic, fire and egress performance. Due to its heritage value it was also necessary to preserve existing exterior and interior architectural finishes and features. Seismic isolation was the solution with least impact upon the existing fabric. As well as attending to preservation and refurbishment, Tadao Ando Architect and Associates introduced several unmistakably modern glazed volumes that contrast with the heavy existing construction. These include the new glazed entry and a light-filled multi-storey rear extension between two concrete cores (Figures 7.6 and 7.7). These additions not only satisfy

7.6 The International Library of Children's Literature, Tokyo, Japan. The main façade with the new glazed entry

7.7 The new rear façade of the library that comprises two reinforced concrete cores and glazed walls

7.8 A circulation and gathering space between the rear existing façade and the new glazed wall

seismic and fire requirements but create attractive spaces contributing to improved building and functionality.

All existing and new construction is isolated by natural rubber bearings and separate lead dampers. The isolation plane generally is located below the ground floor beams but also steps down under the bases of the two concrete cores (Figure 7.9).[5] The construction details are typical of retrofitted load-bearing wall isolated buildings as discussed above. Due to the two different structural materials and systems in the existing building, and the addition of two concrete cores, a new reinforced concrete ground floor, and slabs at other levels, create strong diaphragms that tie the different structural systems together. The building reacts against earthquake forces as a single unit.

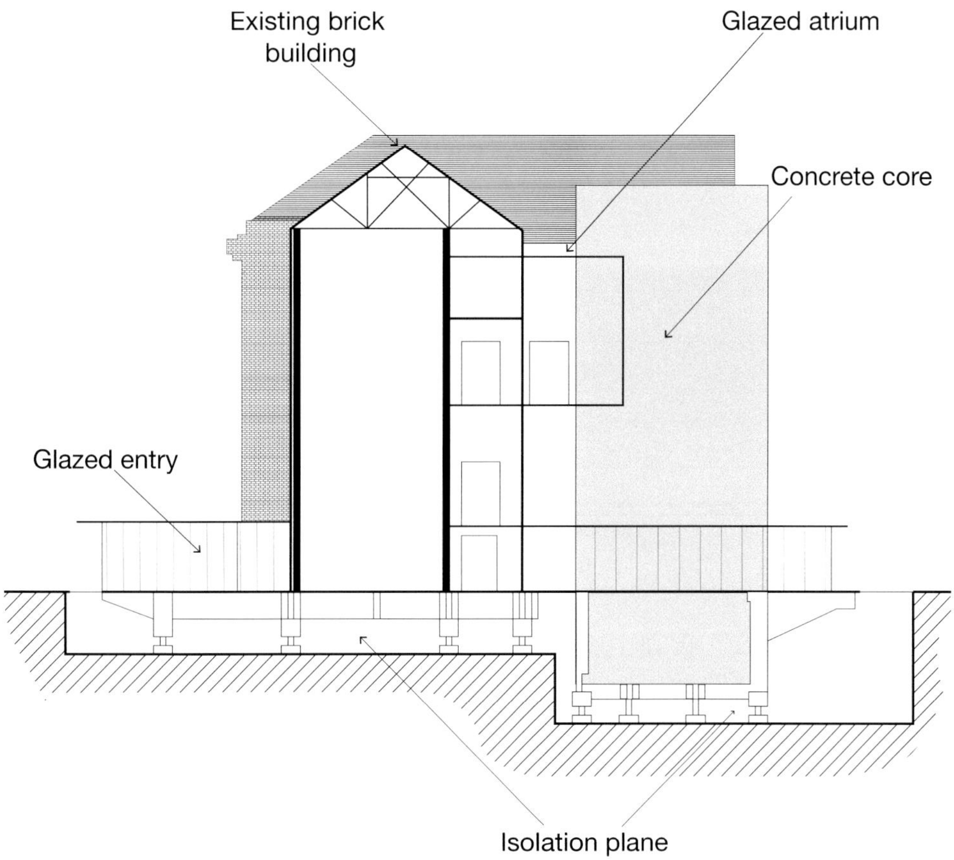

7.9 A simplified section through the retrofitted building showing the varied isolation plane levels

Iasi City Hall, Romania

The historic City Hall was constructed during the 1810s in a neo-classical Viennese style (Figures 7.10 and 7.11).[6] It was damaged during the 1977 Bucharest earthquake, even though the intensity of shaking at Iasi was quite low. Analyses indicated the likelihood of severe damage during the design earthquake, as defined by the current Romanian seismic code. Typical of monumental buildings of its period, it is built of many interconnected unreinforced masonry walls on stone masonry foundations that support the reinforced concrete floor slabs that have replaced the original wooden flooring. The walls are very thick. Exterior walls vary in thickness from 1.0 m to 1.7 m, while interior walls reach 0.6 m thick.

Rather than attempting to retrofit the building using conventional techniques, like inserting new structural elements such as structural walls, it was decided to seismically isolate the building. This solution has avoided the need for any additional superstructure strengthening which would have greatly affected the heritage values of the building. By founding the building on a combination of lead-rubber and sliding bearings, the seismic accelerations experienced in the superstructure walls are reduced enough to prevent damage during the design earthquake.

7.10 Iasi City Hall, Romania

Source: Miyamoto International

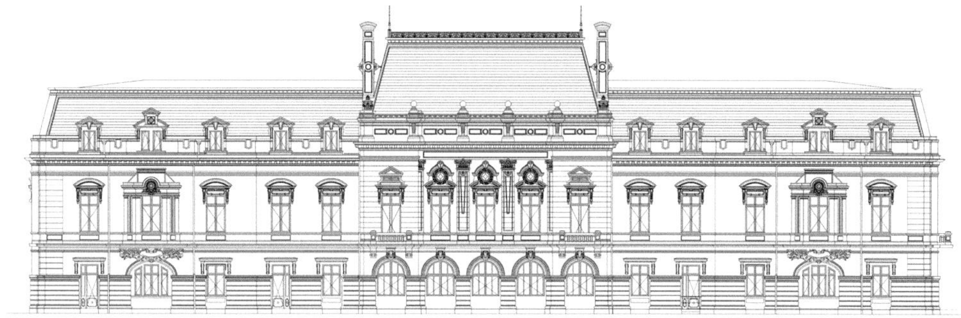

7.11 An elevation of Iasi City Hall

Source: Miyamoto International

Figure 7.12 shows a foundation plan with the many new bearings positioned along walls and at wall intersections. The horizontal isolation plane is just below ground floor level. To distribute gravity forces from the walls into the bearings and then from the bearings into the stone masonry foundation walls below, pairs of permanent shoring (sandwich) beams were placed above and below the isolation plane (Figure 7.13). After casting the beams, which are clamped to the walls by post-tensioned steel rods, openings were made for the bearings and then a horizontal isolation plane gap created along the remaining lengths of wall. A new reinforced concrete diaphragm slab was cast at ground level under the building to tie the foundation walls together and to prevent the new perimeter retaining walls from sliding into the basement space.

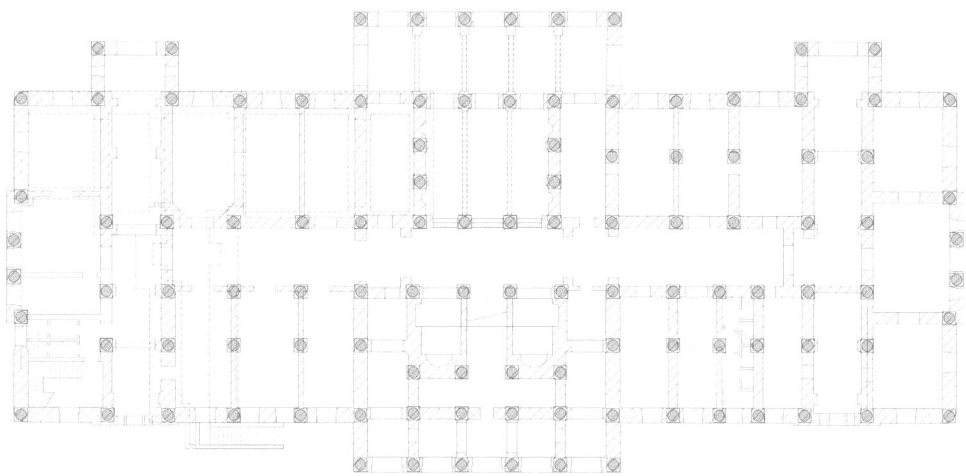

7.12 A foundation plan showing the large numbers of walls and the placement of isolating bearings

Source: Miyamoto International

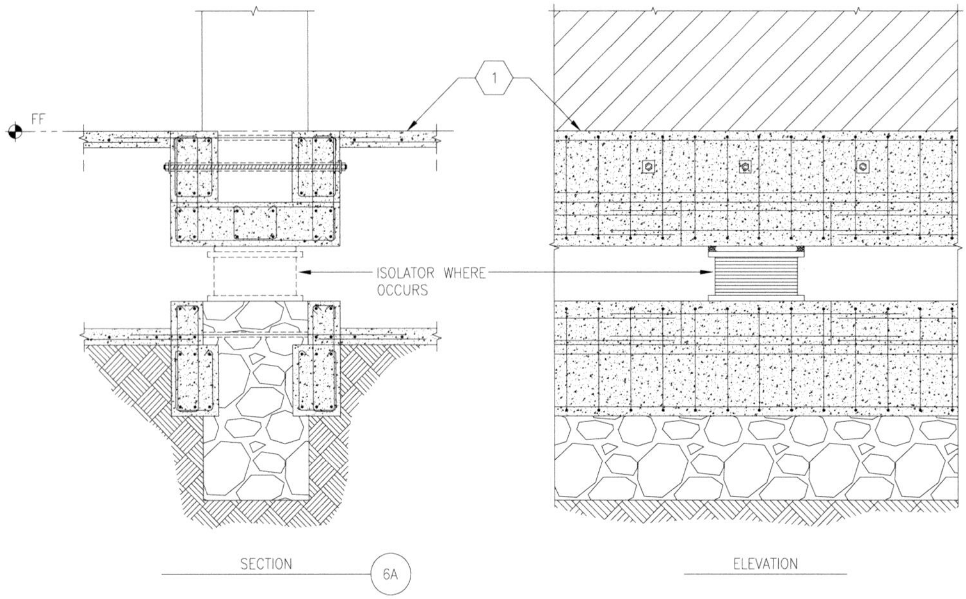

7.13 Details of the pairs of beams required above and below the bearings

Source: Miyamoto International

Case-studies of retrofitting reinforced concrete frame buildings

Rockwell International Building 80

As Ron Mayes explains, there were several noteworthy aspects to this project completed in 1991:

> It was the first base-isolated building retrofit in California, and also the first office building in the USA to be seismically rehabilitated using base isolation while being completely operational. It was also the first framed structure to be retrofitted with base isolators at the mid-height of the bottom-story columns, which avoided disruption to the building occupants during the construction phase.[7]

The original eight-storey building was built in the mid-1960s. Waffle slab floors are supported by reinforced concrete columns and perimeter moment frames. A 1982 structural review of the non-ductile building, confirmed by other reviews five years later, indicated the building possessed only one quarter of the seismic strength required by the then current codes.

> Four retrofit schemes were ultimately selected for analytical comparison: (1) base isolation plus exterior diagonal bracing; (2) conventional diagonal braced frames on the exterior; (3) exterior shear walls in the perimeter frames; and (4) jacketing of the non-ductile concrete beams and columns.[8]

The base isolation plus strengthening scheme was eventually selected. Although it was more expensive than the other schemes it met the owner's directive: 'The building must survive an expected major earthquake with no downtime for the building systems, contents and occupants [. . .]. Structural damage must be limited, and minor in nature. The elevators must resume operation promptly.'[9]

Prior to retrofitting, the recommended exterior diagonal bracing was replaced by stepping reinforced concrete moment frames that far more satisfactorily integrate with the original architecture of the building. In fact, they possibly enhance it. And due to the way they express the need for increased seismic strength towards the base of a building, they are a good example of 'earthquake architecture'.[10]

Normally, the isolating bearings would have been inserted into the basement columns. However, sensitive equipment at that level could not be disturbed, so the isolators were placed in the ground floor. Steel yokes were clamped to the existing columns and hydraulic jacks supported the loads while blocks were cut and removed, and new lead-rubber bearings inserted and fixed in place. The new moment frames were connected to the existing structure by thousands of epoxy-grouted reinforcing bars, the holes of which were drilled at night. Now these moment frames on all four sides of the building above the bearings resist the vast majority of the seismic forces that pass through the base isolation system. The columns beneath the bearings extend down to basement level and are supported by cast-in-place piles.

Since the isolation plane is above ground level it proved a challenge for the two elevator cores to service the basement while passing through the isolation plane and coping with displacements within it. The solution was to support the elevators at first floor level and provide movement gaps at ground and basement level. Vertical support was provided at the base of the elevator pit by Teflon bearings. Now, during an earthquake, the elevators move freely with the superstructure relative to the ground and basement floors.

Rankine Brown Building

This building houses the Victoria University of Wellington library collection (Figure 7.14). It was constructed in the early 1960s in reinforced concrete. Positioned on a sloping site, it has two basement levels over some of its length, a two-storey podium and an eight-storey tower. Its floors consist of precast waffle slabs post-tensioned in two directions through columns to create two-way moment frames. The three ribs that meet the column faces comprise the moment frame beams. There are eight transverse frames and two longitudinal frames, resulting in a total of 16 primary columns (Figure 7.15).

7.14 Rankine Brown Building, Victoria University of Wellington, New Zealand

Prior to retrofitting, a review of the building's seismic vulnerability concluded that after a design-level earthquake demolition would be more economic than repair. The university was also aware of how its whole viability depended on the continuous operation of the library, the possibility of a major earthquake in the Wellington area, and how insurance premiums were increasing sharply.[11] After considering the cost and benefits of different retrofit strategies, seismic isolation was adopted. Its two most attractive features were that no retrofitting work was required above the basement levels, and that the library could continue to operate without significant disruption throughout the duration of the retrofit. The client was also able to negotiate significant earthquake insurance premium deductions that covered the cost of the retrofitting after a relatively short time-frame.

The retrofit project followed the normal steps necessary for isolating an existing frame building. During the project, the lowest basement area was extended the whole length of the building. This gave access to the base of the columns, which were previously part of the foundations, and provided sought-after additional floor space. It meant that the southern area of the podium required temporary support (Figure 7.16). One by one, primary columns were propped to completely remove any compression force at the lowest basement level.

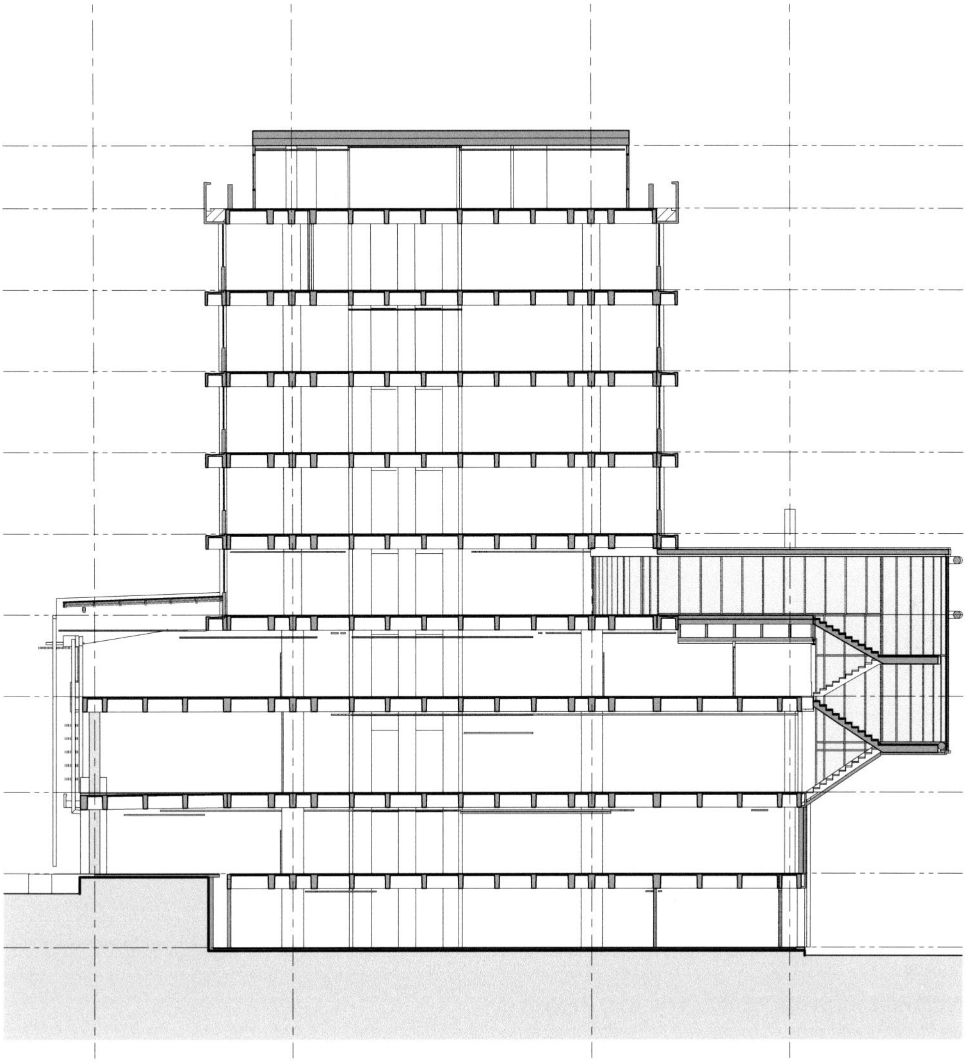

7.15 A cross-section through the Rankine Brown building

Then a wire-saw, with diamonds embedded in the continuous wire, made two cuts through the concrete and the reinforcing steel of the columns. Each cut took approximately four hours. The concrete blocks were removed and new lead-rubber bearings inserted, grouted and bolted to the columns (Figures 7.17 and 7.18). It was fortunate that due to the combination of the basement low inter-storey height and the strength of the existing columns and foundations, these structural elements did not require any additional strengthening.

Lead-rubber bearings were placed only under the 16 primary columns. Smaller concrete columns supporting the podium at the lowest basement level were pinned top and bottom.

7.16 The propping of the podium during the excavation to existing foundation level

They will tilt or rotate when the separated superstructure above them moves up to the maximum displacement of plus or minus 600 mm. Existing floor-to-retaining wall connections were replaced by gaps covered by steel plates able to slide above the retaining walls. At the main entry level these plates are covered by paving stones. Evidence of the isolation system is generally hidden, but in other areas the cover plates are exposed and the isolation plane is visible on close inspection (Figures 7.19 and 7.20).

National Museum of Western Art, Tokyo

Even though it opened as late as 1959 this building is of considerable historic significance, being one of the works of architect Le Corbusier (Figure 7.21). In part, due to the application of some of his 'five points of a new architecture' such as pilotis and the free plan,[12] seismic weaknesses became apparent and were addressed by a seismic retrofitting programme that was completed in 1997. Obviously, it was necessary for any retrofit solution to have minimal impact upon the existing architecture, and so a seismic isolation scheme was adopted.

The original building is of reinforced concrete construction. It consists of three floors above ground and a partial basement. It is square in plan with seven rows of seven columns at 6.3 m centres (Figure 7.22). Unlike the first two examples of retrofitting a frame building, the isolation scheme was implemented by excavating beneath each column to cast a new foundation pad and insert a bearing between the new and existing pads (Figure 7.23). A high-damping rubber bearing is located under each column. Because of seismic isolation only a

7.17 Three props on each side carry the column load as a new bearing is inserted into the gap cut out of the bottom of the column

7.18 A column with a new bearing at its base

7.19 Galvanized steel plates cover the rattle space between retaining wall and isolated superstructure

7.20 The isolation plane, only 20 mm thick, runs along the bottom of the lower black coloured wall band. Vertical sacrificial flashing near the corner of the wall hides the rattle space

small amount of structural improvement was necessary to the superstructure. The ground floor columns were increased in diameter from 530 mm to 600 mm, and those on first floor level required similar strengthening to achieve the desired level of seismic performance. New retaining walls were constructed around the site and adequate separation gaps provided between the main superstructure and adjoining buildings, including the external stairs adjacent to the main façade. The separation details are so well resolved that visitors to the museum have no idea that the building is seismically isolated (Figure 7.24).

7.21 The main entrance to the National Museum of Western Art, Tokyo, Japan

China Basin Landing, San Francisco

This final case-study illustrates a very innovative form of seismic retrofit where a building is retrofitted by adding a two-storey isolated block on top of an existing building.

The client wanted to expand a 1988 three-storey office building but the existing structure could only accommodate a single-storey vertical addition without seismic retrofit. The solution consisted of introducing seismic isolators at the level of the existing roof and then another two storeys of office accommodation was constructed above them (Figures 7.25 and 7.26). These two additional floors above the isolators provide a mass damping effect due to the lower and the isolated storeys vibrating out-of-phase. This sufficiently reduces the seismic demand in the existing structure so that the need for extensive strengthening of the occupied building was avoided. China Basin Landing is the first building in the USA to have a seismically isolated addition on top of an existing building.

7.22 Several interior columns near the perimeter of the building in a gallery

7.23 A cross-sectional model of the isolated museum displayed in the museum. Note the varied levels of the isolation plane

7.24 A seismic separation gap was created between the external stairs and the main building

Limitations

IRRESPECTIVE of the structural systems and materials of an existing building to be seismically retrofitted, construction is always challenging. This is certainly the case with conventional seismic retrofitting where new moment frames, braced frames or structural walls are inserted and connected to the existing superstructure. And the challenge may not be any less when retrofitting by seismic isolation. It's not surprising then that designers and contractors are always trying to develop easier and more economic construction methods. In one new approach developed in Italy, a new bottom foundation concrete slab is cast around the existing foundations. Then another slab, which is strongly connected to the foundations, is cast on top.

7.25 China Basin Landing office building, San Francisco, USA. The existing three-storey 1980s building was retrofitted by placing isolators at roof level and then constructing another two storeys

Source: Dynamic Isolation Systems Inc.

The foundations are cut just below the upper slab which is then jacked up so that isolating bearings can be inserted between the two slabs.[13]

Although a retrofitting scheme comprising seismic isolation may be the preferred method of retrofitting, it may not be feasible due to the need for wide seismic separation gaps between the retrofitted building and its neighbours. If a building intended to be seismically isolated is built close to or against its site boundary then seismic isolation is probably not a feasible solution. Imagine how unrealistic it would be to demolish say a 400 mm wide perimeter strip from the floor plates of a building. The task would be daunting even for a building like the National Museum of Western Art with its perimeter columns set in from the façades. The need for wide separation gaps between adjacent buildings is therefore one of the most common factors preventing widespread adoption of seismic isolation as a seismic retrofitting strategy.

7.26 China Basin Landing office building during construction

Source: Dynamic Isolation Systems Inc.

Notes

1 Allen, E.W. & Bailey, J.S., 1988, 'Seismic rehabilitation of the Salt Lake City & County building using base isolation', Proceedings of the 9th World Conference on Earthquake Engineering, Tokyo-Kyoto, vol. 5, paper 7-9-5, pp. 639–44.

2 Masuzawa, Y. & Hisada., Y. 2012, 'Current state of retrofitting buildings by seismic isolation in Japan', Proceedings of the 15th World Conference on Earthquake Engineering, Lisbon, p. 3.

3 Ibid. p. 4

4 National Diet Library 2006, 'Architecture of the International Library of Children's Literature: preservation and renovation of an old brick building from the 1900s', Conference of Directors of National Libraries in Asia and Oceania (CDNLAO) Newsletter, no. 55, p. 1.

5 Tasaka, M., Mori, N., Yamamoto, H. *et al.*, 2008, 'Applying seismic isolation to buildings in Japan – retrofitting and middle-story isolation'. Proceedings of the 18th Analysis and Computation Speciality Conference, ASCE, Vancouver, p. 4.

6 Glilani, A.S. & Miyamoto, K.H., 2102, 'Base isolation retrofit challenges in a historical monumental building in Romania', Proceedings of the 15th World Conference on Earthquake Engineering, Lisbon, p. 1.

7 Mayes, R.L., 2007, 'R.L. Mayes', *Structural Design of Tall and Special Buildings*, vol. 16, no. 1, pp. 3–36, p. 22.

8 Gates, W.E., Nester, M.R. & Whitby, T.R., 1992, 'Managing seismic risk: a case history of seismic retrofit for a non-ductile reinforced concrete frame high rise office building', Proceedings of the 10th World Conference on Earthquake Engineering, Rotterdam: Balkema, pp. 5261–6, p. 5263.

9 Dooley, C.T. & Robison, R., 1990, 'Seismic surgery', *Civil Engineering, ASCE*, vol. 60 no. 9, pp. 72–5.

10 Charleson, A.W., 2008, *Seismic design for architects: outwitting the quake*, Amsterdam: Elsevier, p. 251.

11 Clark, W.D. & Mason, J.E., 2004, 'Base isolation of an existing 10-storey building to enhance earthquake resistance', Proceedings of the New Zealand Society for Earthquake Engineering Technical Conference, Rotorua, paper 10, p. 1.

12 The National Museum of Western Art, 'Discover Architecture Map'. Available from: www.nmwa.go.jp/jp/about/pdf/discoverarchitecturemap_en.pdf (accessed: 22 September 2014).

13 Briseghella, B., Zordan, T., Zambianchi, L. *et al.*, 2012, 'Lift-up and base isolation as a retrofit technique for R.C. existing building', Proceedings of the 15th World Conference on Earthquake Engineering, Lisbon.

– 8 –

Design aspects

Introduction

THIS chapter focusses upon seismically isolated buildings with an emphasis upon their unique design requirements as compared to the design of conventional buildings. We assume that architects are responsible for architectural form which is developed in collaboration with structural engineers for both types of buildings. After discussing the special requirements for the composition of and the effectiveness of the design team, such as the need for more intensive communication and collaboration, we turn our attention to how design for seismic isolation impacts structural engineers. Then, in the major section of the chapter, we discuss the full range of architectural design implications of a seismically isolated building. These cover conceptual architectural design to construction detailing.

Design team

THE successful introduction of any new technology into a design and construction project necessitates additional communication and a higher level of collaboration between members of the design team. Seismic isolation is no exception. Even before making a final decision to isolate, numerous meetings and briefing sessions involving client, architect and structural engineer are to be expected. In the design of a conventional fixed-base building, codes of practice are followed to ensure a 'code compliant' design. Implicit assumptions embedded in codes regarding seismic risk and performance are usually accepted uncritically by the design team. However, if seismic isolation is a possibility, engineers need to brief the architect and the client on matters that are mainly only discussed by committee members developing seismic-related codes.

Once a decision to seismically isolate is made, the intensity of collaboration increases. Seismic isolation brings with it additional design and detailing issues needing resolution. For example, architect and structural engineer together have to decide how and where to isolate.

Will the isolation plane be above ground or within a basement? Then, after running numerous dynamic analyses, the engineer needs to provide information about the maximum horizontal displacements along the isolation plane to be designed and detailed for. This information is additional to values of superstructure inter-storey deflections that are always needed to provide adequate clearances between structure, such as columns, and non-structural components, such as partition walls. Estimates of residual displacements of the isolation system after a major earthquake, which may affect various details such as cover plates over seismic separation gaps, also need to be determined by the engineer and communicated to the architect. All this additional complexity requires greater interaction within the design team, as well as determination within it to work together for the sake of a successful project.

Structural engineering design

ONE of the challenges facing structural engineers designing seismically isolated buildings is a lack of definitive, yet what are considered as overly, restrictive design codes. Although some countries, such as Japan, have developed robust and streamlined design protocols, in the USA, New Zealand and Europe, for example, significant improvements to existing code requirements are only just being completed, new rules developed and further changes expected.[1,2,3] These are concerted efforts to reduce regulatory barriers, introduce consistency of design and, at least for reasonably regular buildings, to simplify design procedures.

Although these in-country or regional developments will undoubtedly remove some barriers to seismic isolation, considerable international variance regarding aspects of seismic isolation design remain. Alessandro Martelli states that 'SI [seismic isolation] is considered as an additional safety measure (with consequent additional construction costs) in some countries (Japan, USA), while in others (including Italy) the codes allow to partly take into account the reduction of the seismic force acting in the superstructure'.[4] As another example, in Japan the design earthquake is approximately a 500 year event, but in the USA isolator displacement is based on a 2500 year return period.[5] Peng Pan points out that in Japan a factor of safety of 1.5 is used to determine the horizontal isolation clearance dimension[6] but designers in the USA are not required to use a factor of safety, possibly because of the larger design earthquake. Some designers from Japan, China, USA, Italy and Taiwan are attempting to standardize design procedures internationally,[7] but it is too early to rate their chance of success.

As well as the challenges described above, the structural design of a seismically isolated building necessitates considerable additional engineering input including:

- meeting special structural design requirements and using sophisticated, infrequently used and time-consuming design tools such as inelastic time-history analyses;

- participation in independent design reviews;
- designing additional components such as the bearings, their fixings, suspended lift-pits, movement and separation details, and retaining (moat) walls;
- interaction with specialist consultants such as seismologists, geotechnical engineers and bearing manufacturers;
- greater involvement with services engineers and architects in the isolation plane area;
- reviewing bearing performance test results and possibly observing bearing tests; and
- on-going monitoring of the isolation system as part of its maintenance schedule (see Chapter 10).

Due to the extra work undertaken by the structural engineer it is reasonable to expect a commensurate increase in professional fees.

One of the most important items of information the structural engineer provides to the architect is the width of the seismic or movement gap. Its purpose is to allow horizontal movement between the isolated superstructure and the foundations, retaining structures or adjacent buildings above the level of isolation plane. Cutfield and Ma conclude their cost-benefit study of an isolated and a conventional building by noting:

> Overall, the performance of the isolated building models was far superior to the conventional building model; however, this performance degraded somewhat in the unlikely event of structural pounding against the building's moat wall. Hence, considered and conservative selection of the building seismic gap is important for achieving the best performance from a base isolated building.[8]

There is no doubt that pounding is potentially dangerous, especially if the superstructure is not ductile. This is the reason proposed changes to ASCE 7–10[9] require that the moat clearance be increased by 20 per cent for non-ductile structures, such as ordinary concentrically braced steel frames.

A conservative approach to determining the movement gap width appears wise for several other reasons as well. Movements calculated for some soil sites subject to bi-directional near-fault ground motions are up to 15 per cent greater than expected,[10] and movement that increases at corners of buildings due to torsion can reach up to 30 per cent if buildings are long.[11] Wider movement gaps may also assist in future-proofing. For example, the 150 mm gap around the William Clayton Building that was constructed in 1981 (Figure 2.1), although based upon the best seismological advice of the day, is now considered too narrow.

However, overly wide movement gaps are not necessarily the answer. Although moat walls might be a source of pounding, they can also prevent elastomeric bearing instability. Wide gaps also mean greater clearance around the perimeter of the building. This might reduce useable floor area and increase the possibly undesirable visual impact of architectural details covering the gaps.

Architectural design

SEISMIC isolation increases the architect's as well as the structural engineer's workload. At both conceptual and final detailing levels of design, many design scenarios require addressing that are irrelevant for conventional buildings. Yet, while detailing for the large movements across an isolation plane can be challenging, the significantly reduced movements within an isolated superstructure can simplify detailing elsewhere.

As mentioned at the start of this chapter, early collaboration between architect and engineer is especially important for the design of a seismically isolated building. Many conceptual aspects of the isolation scheme require discussion and resolution in those first few design team meetings. The decision regarding the height of the isolation plane has already been mentioned, but there are many others. For example, should a bearing be placed under each column, or are there advantages in providing transfer beams to reduce the numbers of bearings? How might that decision help or hinder the provision of the minimum recommended 1.2 m crawl space height for inspection and maintenance underneath the beams protruding into that space (Figure 8.1). Or, if there are several multi-storey blocks to be isolated, should they have separate isolation systems, or might founding them on a common isolated slab be a better solution (Figure 8.2)? Also at the conceptual stage of design, preliminary ideas of key details benefit from a collaborative approach. Strategies for achieving the horizontal clearances for movement along the isolation plane and up the perimeter of a building need resolution, and these will ultimately affect the built floor area (Figures 8.3 to 8.5).

In Japan, the recommended clearance to protect mainenance staff from being crushed during an earthquake is 200 mm wider than the design displacement unless the area immediately adjacent to the structure is a walkway. In this case the additional clearance should be 800 mm to protect the public. Signage is sometimes provided to warn people of the danger of the gap narrowing during an earthquake.[12]

Later in the design process detailed design is undertaken, but before considering typical architectural details, observations after the 2011 Tōhoku earthquake, Japan, indicate that their performance left a lot to be desired. Although in principle the provision for unrestrained and damage-free movement is straightforward, it is more difficult to achieve in practice. Saiki and others' survey of over 300 isolated buildings revealed that 30 per cent of the buildings experienced damage to movement joints.[13] Even though the movements were in most cases far less than the maximum design displacements, joints did not function as intended by their designers. Defects were observed due to the location of joints, obstructions in their immediate vicinity and lack of maintenance. The authors suggest a pragmatic and sensible approach to specifying adequate performance of movement details. They accept minor damage in locations other than evacuation routes and acknowledge different degrees of damage are appropriate for different earthquake intensities. After defining four levels of damage (Table 8.1), they propose movement joints are classified and specified by a three level Performance Index (Table 8.2). These tables are based on those proposed in their paper.

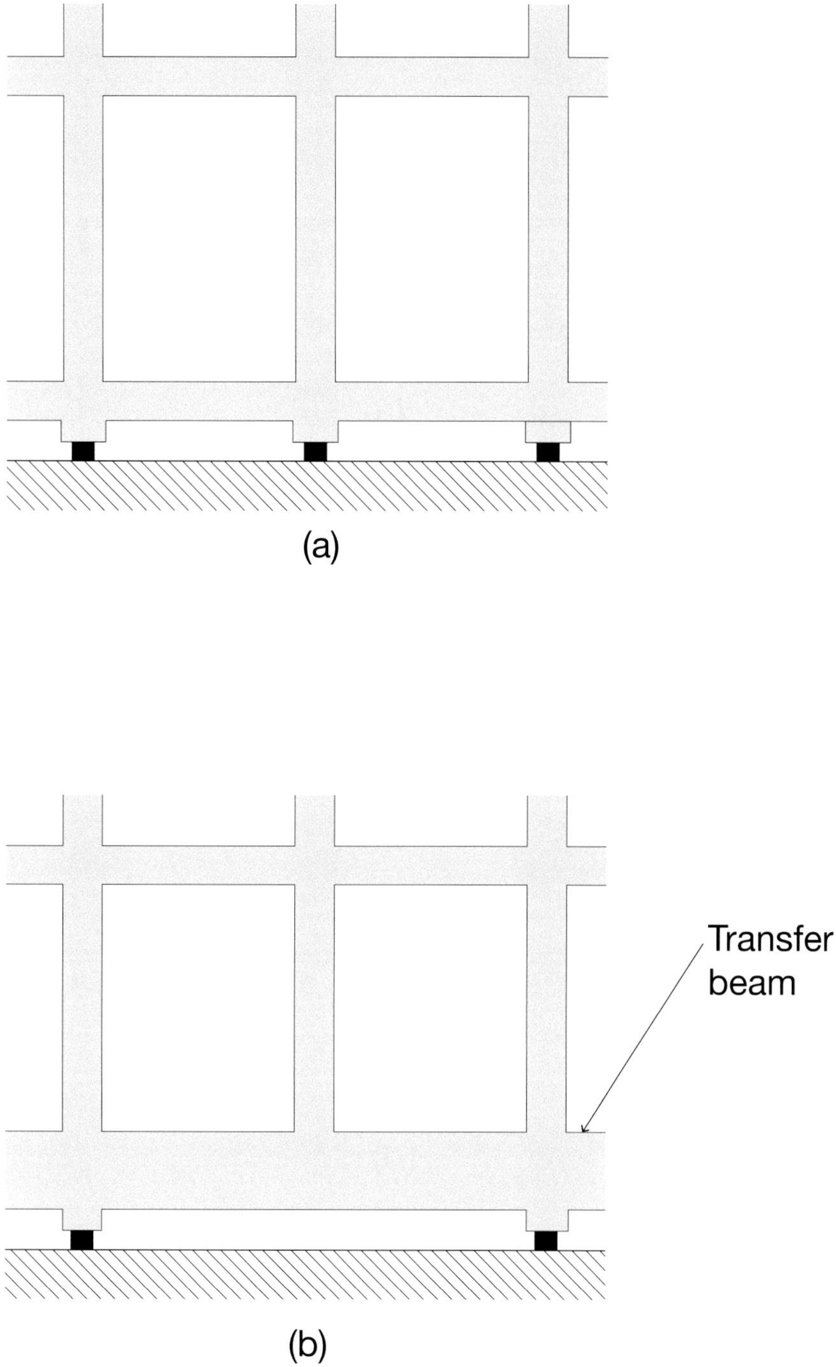

8.1 A bearing is usually placed under each column (a), but a transfer beam reduces the numbers of bearings while maintaining a crawl space (b)

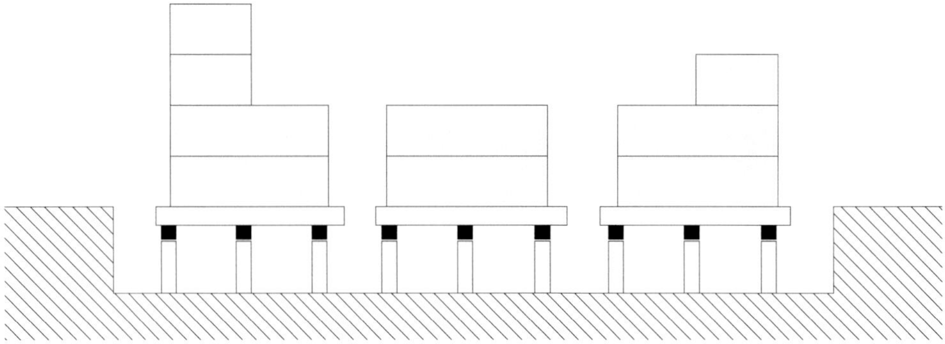

(a)

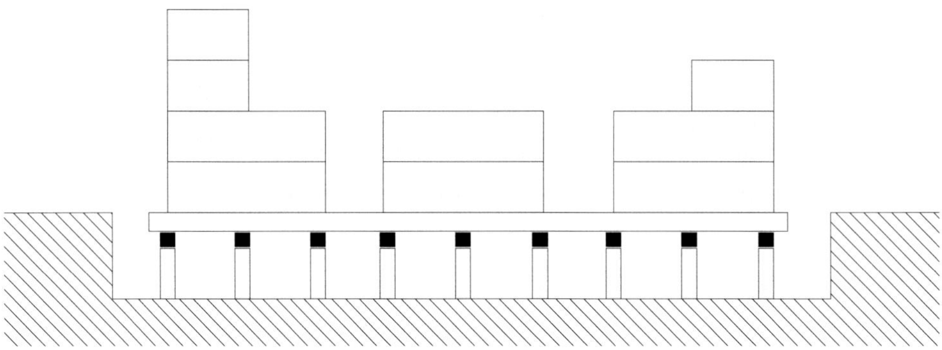

(b)

8.2 Several separate buildings can be individually isolated (a), or share a single isolated slab (b)

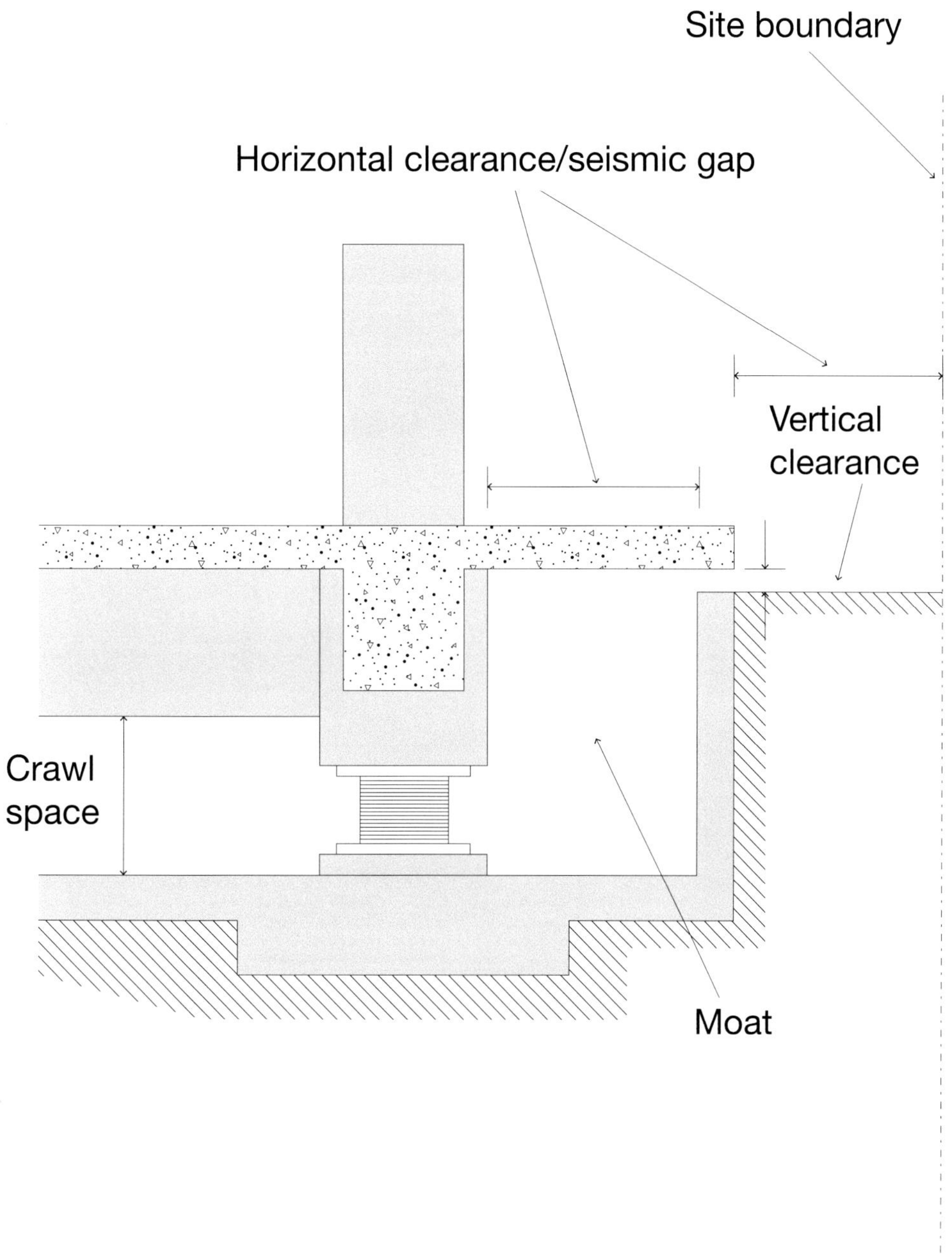

8.3 Clearances required around an isolated building

8.4 Clearances either side of an isolated building, Tokyo, Japan

8.5 Horizontal clearance between a boundary retaining wall to the left and an isolated building to the right, Tokyo, Japan

Table 8.1 Definition of damage to movement joints

Damage category	*Description of damage*
Fully functional	No deformation, change of slope or opening of a gap that affects functionality. The joint can be used continuously without repair. Minor damage such as scratches to finishes or cuts to seals are acceptable.
1	No excessive damage due to deformation, change of slope or gaps. Some adjustments and repairs may be required but functionality is unaffected. All areas are accessible even though there may be some differences in level or protrusion of wall elements.
2	Significant damage affects but does not prevent function. Large scale repair or replacement of joints is necessary, but no elements are detached making some floor areas inaccessible.
Loss of function	Major damage leading to loss of function. Continuous use immediately after an earthquake is not possible.

Table 8.2 Categories of movement joint performance for two earthquake intensities, different joint locations and performance verification methods

Performance index category	*Damage category at low intensity shaking (up to 100 mm movement)*	*Damage category at high intensity shaking (up to the maximum design movement)*	*Location of movement joint*	*Verification*
A	Fully functional	Fully functional	Evacuation routes, high traffic people and cars	Dynamic movement test up to maximum design movement
B	Fully functional	Category 1	Accessible areas	Dynamic or a more simple movement test, possibly by hand
C	Category 1	Category 2	Minimal access by people	Review of working drawings

Note that the functionality implicit in these tables relates to human use. Designers also need to check that other aspects of building function such as fire protection, weather and acoustic proofing perform as intended after movement joints are displaced, and especially if there are any permanent off-sets. For long-period isolated buildings (periods equal to or greater than 4.0 seconds) and with high bearing yield values, residual displacements of 25 to 150 mm may occur.[14] Structural engineering advice should be sought.

The remainder of this chapter focusses upon specific details of seismically isolated buildings that architects design and detail to allow movement relative to the ground. Sketches of typical details are supplemented by images of examples from around the world.

Moat area (rattle-space) and horizontal cover plates

These details apply around the perimeter of seismically isolated buildings. As their names suggest, they allow unrestrained movement of the isolated superstructure relative to the ground in *any* direction. As shown in Figures 8.6 and 8.7, the main requirement is to achieve the primary horizontal movement gap and then to cover it, to allow access in some areas, and to prevent ingress of water, rubbish or snow. The width of the horizontal gap, which is provided by the structural engineer, is dependent upon many factors. Typically, its width is in the order of 300–500 mm, but can become as wide as 900 mm.[15] The moat cover can consist of a cantilever slab, hinged slabs or steel plates. It usually connects to the isolated structure and rests on the top of the retaining wall, simply sliding along it and over it (Figures 8.8 to 8.10). Steel plates are particularly common moat covers at entrances of buildings and across vehicle ramps (Figures 8.11 to 8.13).

There are many instances where moat covers or movement joints abut exterior paving where the level of paving and the surface of the isolated ground floor align. In these cases a typical detail employs angled sliding surfaces that prevent compression occurring in the cover plate when the building moves towards the paving (Figures 8.14 to 8.18). Several companies manufacture and supply cover plates for floors, both exterior and interior. Information published on the internet can be consulted. The wide range of cover plates includes those that self-centre (Figures 8.19 and 8.20). When installed correctly and paved, they can be almost invisible (Figure 8.21), although it can be convenient to use cover plates that allow water to enter a drain that is incorporated into or near a moat (Figure 8.22). Internal cover plates utilize the same principles as those discussed above. They are also available in many different forms, ranging from simple highly visible surface-mounted plates to those that are more concealed (Figures 8.23 and 8.24). They may incorporate additional complexity to provide fire resistance across the gap.

In some projects the expense of specifying relatively sophisticated and expensive cover plates may be queried due to the low risk of damage and the probable ease of post-earthquake repair. Some clients might find it acceptable to use very simple cover plates with minimal

8.6 Horizontal clearance between an external stair to the left and an isolated building to the right, Tokyo, Japan

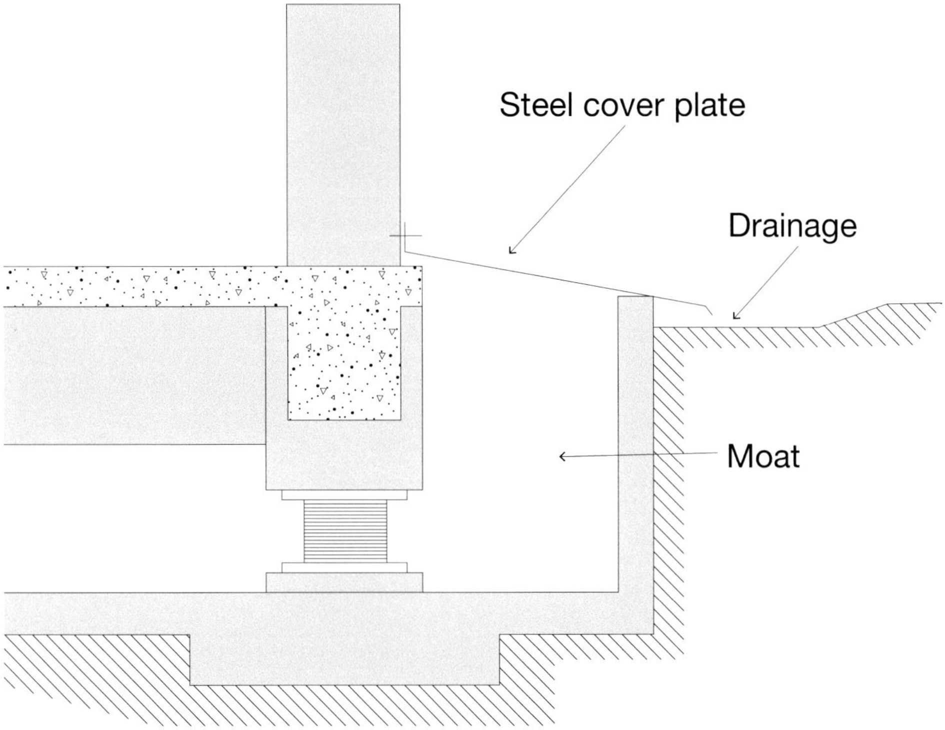

8.7 A section through a typical moat, a perimeter column and bearing where the moat is covered by a cover plate connected to the isolated structure and able to slide on the retaining wall beneath it

movement allowance, but protected against falling with restraint wires in the event of large displacements (Figure 8.25).

Other movement joint coverings

As well as moat covers and other flooring movement joints, capacity for seismic movement is also required where an isolated building connects to another building, be it isolated or not, at the roof, up walls and along and across ceilings. Just as proprietary covers are available for moat and floor seismic gaps, so it is with ceilings, roofs and walls. However, it is relatively easy to create movement joints for ceilings using negative joints, and allowing ceiling planes at different levels to slide along and across each other (refer to Figure 8.25). Movement gaps at roof level can be accommodated by using a 'first-principles' approach. A solid cover is attached on one side only, and then covered by some type of flexible metal flashing. Wall joints are more difficult. While allowing for inwards and outwards movement in the plane of a wall is

8.8 A cantilever slab moat cover that can move into and away from a garden area, Tokyo, Japan

8.9 A cantilever slab moat cover moving into and away from a garden on one side and forming a step at a paved area on the other, Tokyo, Japan

8.10 A steel plate moat cover attached to the isolated building on the right and able to slide over the retaining wall and paving to the left, Wellington, New Zealand

straightforward, displacements normal to the walls are far more difficult to achieve while simultaneously coping with wind pressure and other weather-proofing requirements (Figure 8.26). One approach is to specify a complex proprietary system consisting of magnets, springs and a hinge which would not be expected to be damaged even during extreme movement (Figure 8.27). Alternatively, small movements may be allowed for by minimal movement provision and by elastic bending of wall linings and some members designed to be sacrificial in a large event, perhaps sacrificial steel work and frangible lining panels. This was the detailing philosophy adopted for the connection between the buildings shown in Figure 8.28.

The challenge of designing movement joints that meet performance objectives acceptable to the client should not be underestimated. It is likely that several iterations of

8.11 At a main entrance the cover plate is attached to the isolated superstructure on the left and able to slide over the paving on the right. The fixings of the vertical steel plate to the left of the handrail are designed not to impede the movement of the stair towards it, San Francisco, USA

design development will be required. Designers should consider constructing small physical models to both develop the design and help communicate it, possibly to the client, but definitely to the contractor. Assume that a contractor has never had to allow for such large seismic movements before.

Other movement details

Detailing for movement is necessary whenever a building element crosses an isolation plane. Stairs often fall into this category. To prevent them reducing the effectiveness of the seismic isolation system and to avoid damage to the stairs themselves during seismic movement, separation is needed. Both the stair structure and handrails require separation (Figures 8.29 and 8.30).

Elevator shafts often cross isolation planes. The usual strategy is to fix the elevator shaft to the isolated structure and suspend it so that the isolation plane passes beneath it.

8.12 A steel cover plate at a secondary entrance slides over the exterior concrete paving. The chain linking the building to the handrail is considered sacrificial. It will break and be replaced after an earthquake, Wanganui, New Zealand

8.13 A steel moat cover across a vehicle ramp, Tokyo, Japan

8.14 A transition area where the surface of a cantilever moat cover that moves above the surrounding paving gradually becomes level with that paving, Tokyo, Japan

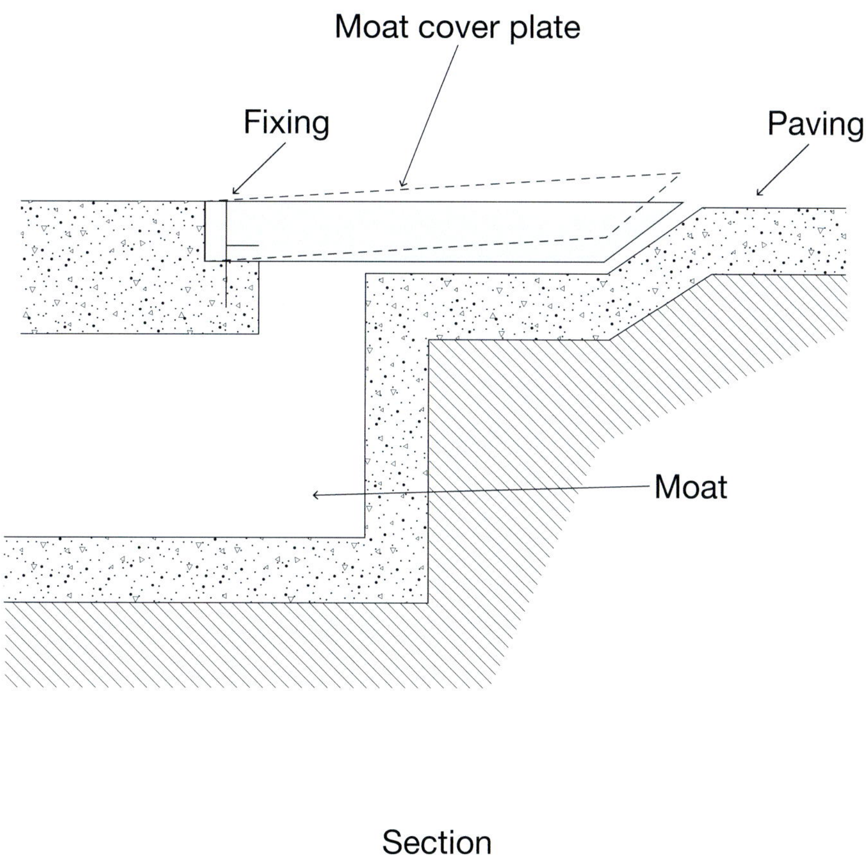

8.15 A detail of an angled sliding interface between a moat cover and paving to avoid the paving preventing movement of the isolated structure to the right

8.16 45 degree sliding surfaces between concrete paving and a steel grating cover plate, Wellington, New Zealand

8.17 Detailing of a movement joint between isolated and fixed paving that utilizes angled sliding planes, Tokyo, Japan

8.18 A precast moat cover where movement parallel to the entrance is accommodated by the nearer angled sliding joint. Movement at right angles occurs when the precast slabs slide along the steel support rails that are attached to the isolated structure, Tokyo, Japan

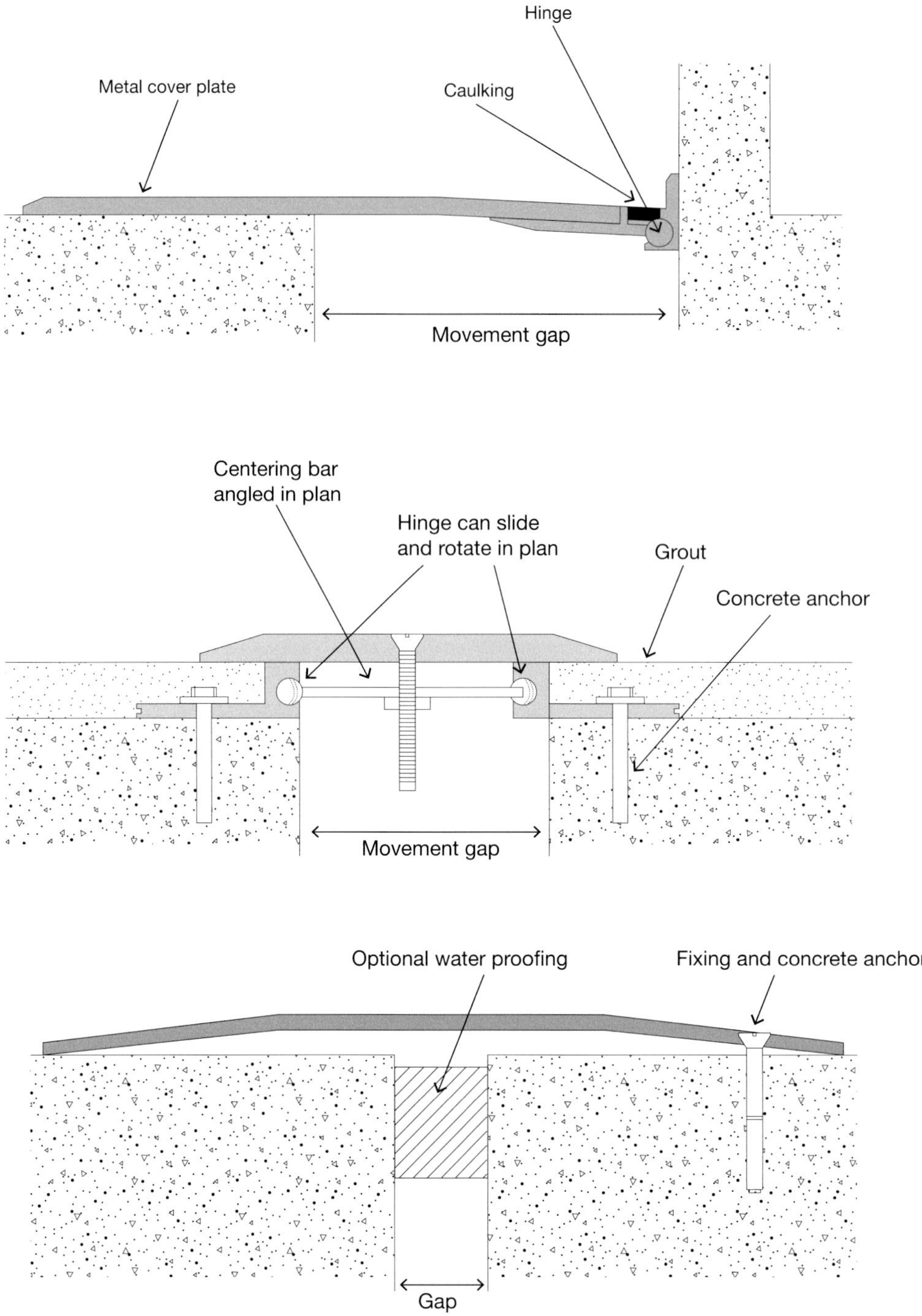

8.19 Three types of cover plates. All plates allow horizontal movement in any direction

8.20 The underside of a self-centring floor cover plate. Diagonal bars that centre and support the plate are visible as well as a flexible elastomeric wall joint, Wanganui, New Zealand

8.21 A cover plate allowing a full range of movement is almost indistinguishable from the surrounding area of paving. The plate is fixed to the building and due to its 45 degree sliding surface will slide across and along the moat gap below, Tokyo, Japan

The elevator shaft needs to be braced so that it can resist horizontal inertia forces as it cantilevers from the floor above (Figure 8.31). A cover plate over the clearance gap must be provided at elevator doors and where otherwise visible.

If the isolation plane is located mid-storey then the partition walls of that storey must be detailed for movement. As shown in Figure 8.32, movement along the isolation plane must not be obstructed, so the walls require all their structural support from (usually) the floor below. If large movements are to be accommodated without damaging the wall, it should be physically separated from the floor above. Its strength against wind pressure and other horizontal forces can be provided by cantilevering it from its base using specifically designed vertical wall structure rigidly fixed to the floor beneath, or else using return walls to brace it. The gap at the top may need to be filled with soft fire-proof material and covered by a scotia which may be sacrificial. If some wall damage is acceptable the wall need not cantilever but can be fixed normally at its base and connected to the floor soffit or beam above in a way that allows sliding movement along its length.

8.22 A steel grill cover plate where drainage is integrated with a moat space, Tokyo, Japan

Although the architect will not personally detail movement capacity of pipes and wiring crossing an isolation plane, some extra provision of space to accommodate it may be necessary. The principle is straightforward. Pipes need to be flexible or pin jointed to undergo the design displacement without damage (Figures 8.33 to 8.35). Wiring should be provided with sufficient slack or excess length. Because flexible or multi-pinned pipe joints are expensive, cost savings are possible by connecting pipes together in the isolated superstructure in order to minimise the numbers of pipes crossing the isolation plane.

8.23 A visible surface (and wall) mounted interior cover plate, San Francisco, USA

8.24 A flush mounted interior floor plate, Wanganui, New Zealand

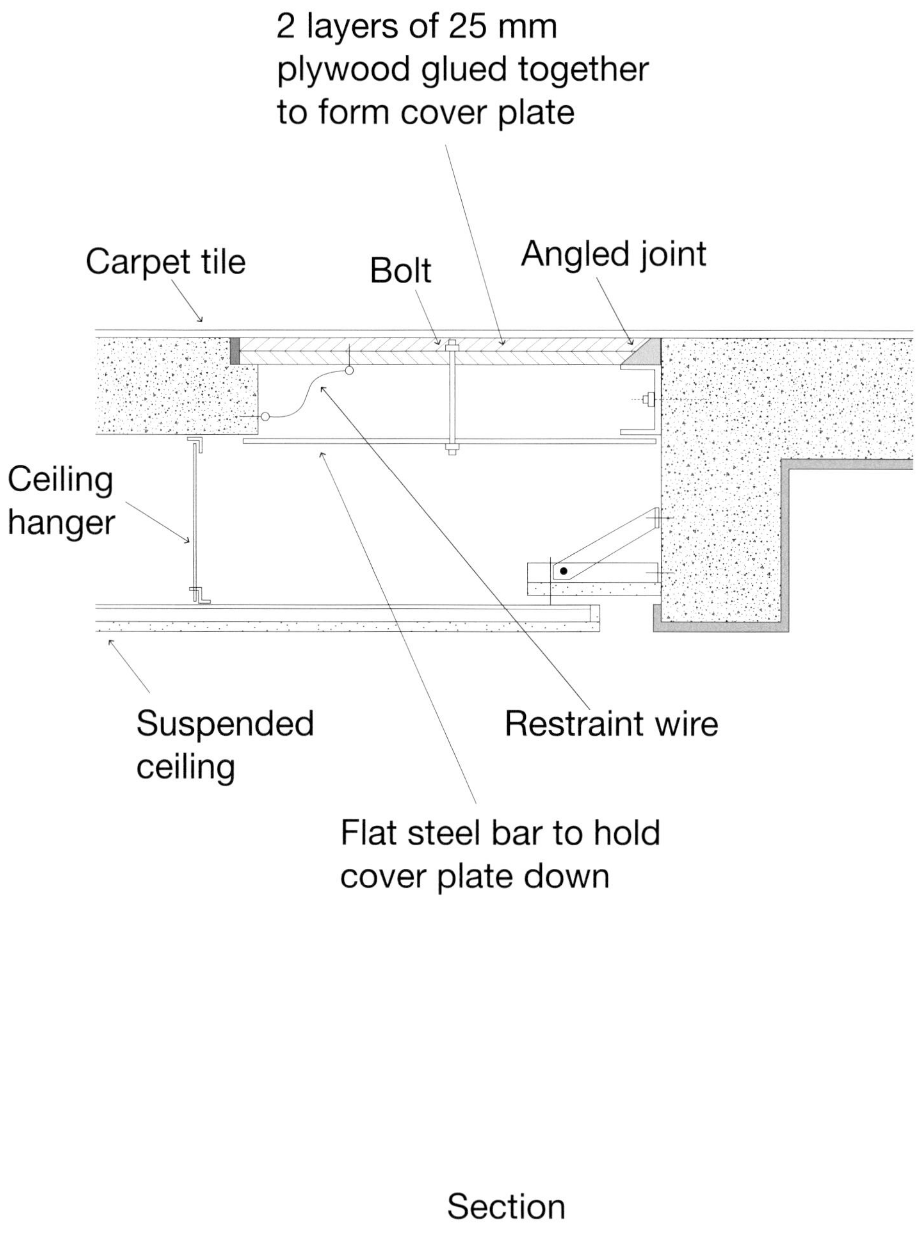

8.25 An example of a low-cost floor cover plate which will need repair following a moderate earthquake. Movement at ceiling level is accommodated by a negative detail

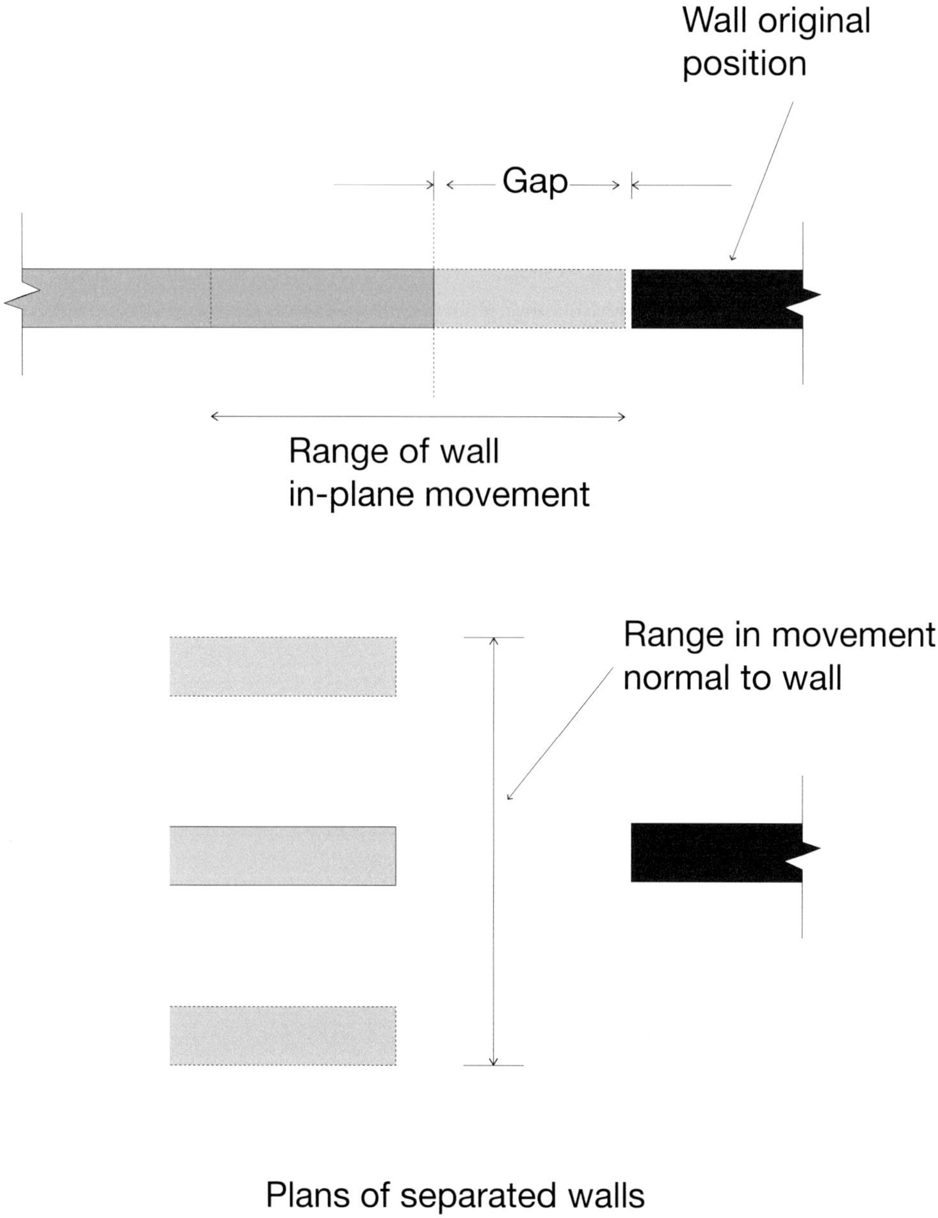

8.26 Accommodating movement of two walls towards and away from each other is easily achievable, but movement normal to them is more challenging

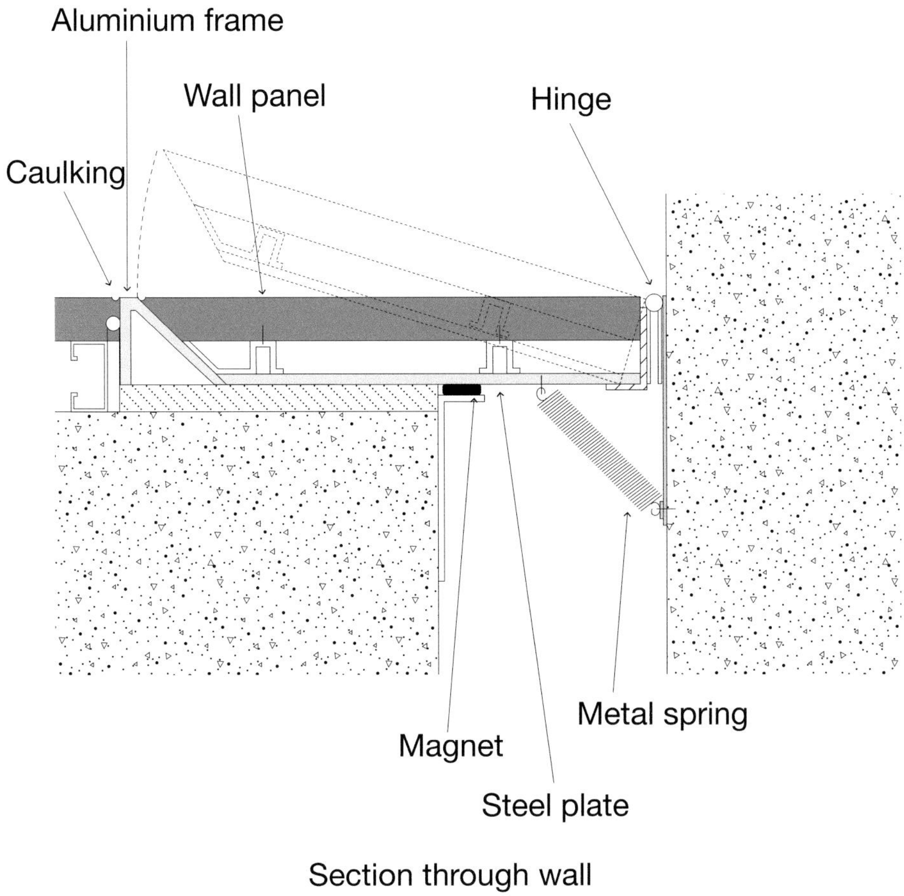

8.27 A sophisticated wall movement joint that allows the gap width to change, movement along the plate and the two walls to move relative to each other at right angles to the wall panel

8.28 The vertical black painted thin metal covering is part of the wall movement joint between the seismically isolated building to the left and the fixed-base building to the right. The choice of materials and colour achieve an elegant transition. Sacrificial steel elements and frangible interior linings will need replacement after a moderate to large earthquake, Wellington, New Zealand

8.29 The stair stringer and balusters are cut to allow movement in all horizontal directions. The handrail is cut (not shown) and detailed in a similar manner, Wellington, New Zealand

8.30 An external secondary stair. The stair landing cantilevers from the isolated building to the right and slides on the supporting pier. The handrails are separated to allow movement, Lower Hutt, New Zealand

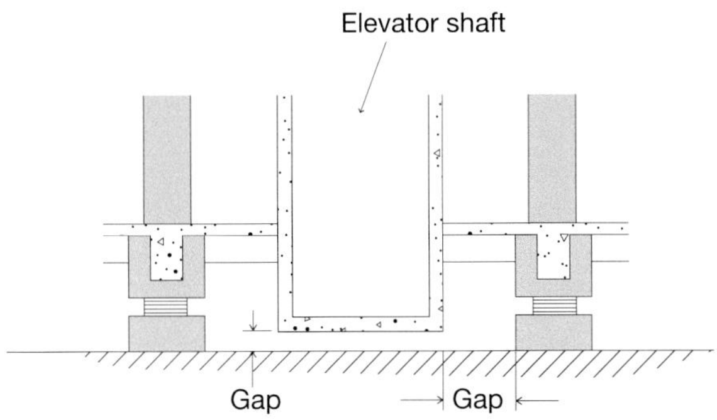

(a)

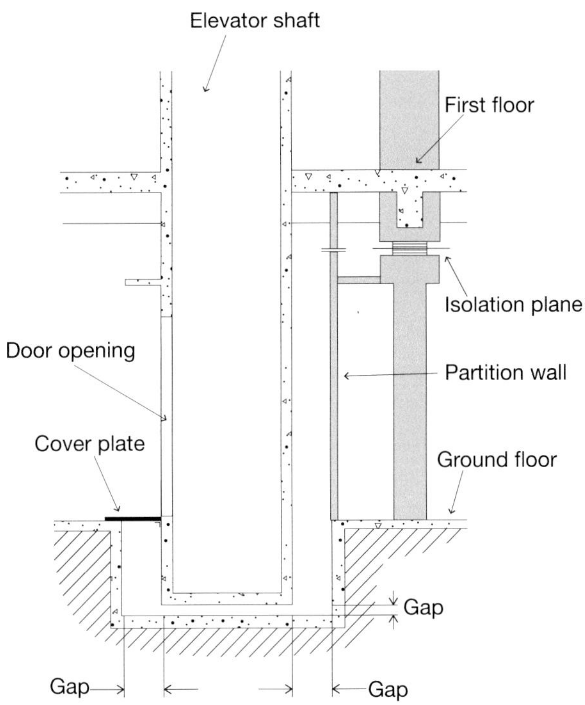

(b)

8.31 Sections through an elevator shaft showing how it hangs from the superstructure and is braced back to it. It is separated from any other elements that might prevent its movement with the isolated superstructure. In (a) the isolation plane is at the base of the building, while in (b) the isolation plane is nearer mid-storey height

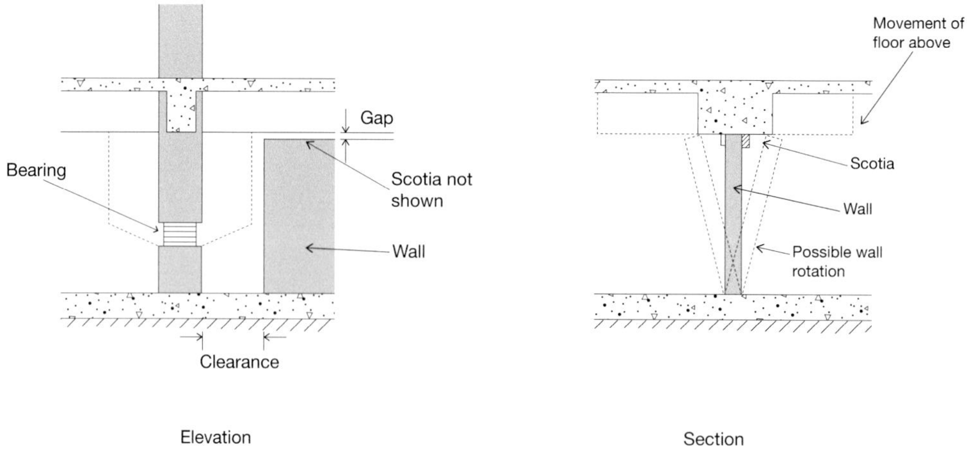

8.32 The need to provide movement between walls and upper floor where the isolation plane is between floors. If a wall is self-supporting normal to its length, then scotias must be weakly attached to allow movement and be reinstated after an earthquake. If the wall relies on its connection to the structure above for its stability, the scotias must provide that restraint

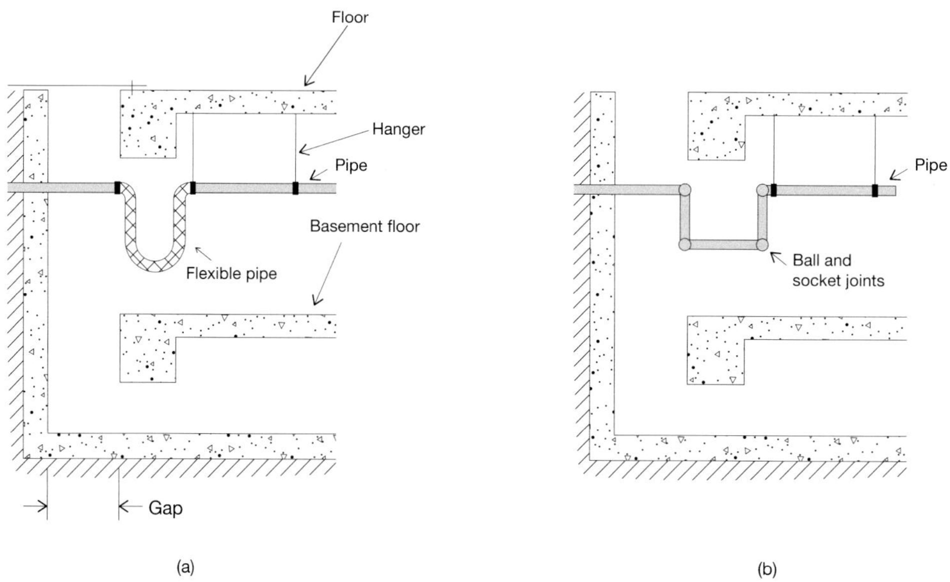

8.33 Pipework needs flexibility (a), or several short pin-jointed lengths to allow damage-free seismic displacement (b)

8.34 Movement in a pipe connecting an isolated and fixed-base building is achieved by several pipe lengths with ball and socket joints that allow rotation between pipes, San Francisco, USA

8.35 A flexible pipe crossing an isolation plane, Tokyo, Japan

Notes

1 *ASCE 7 2009: Minimum design loads for buildings and other structures*, American Society of Civil Engineers, Reston VA., Chapter 17 Seismic design requirements for seismically isolated structures.

2 Booth, E., 2014, *Earthquake design practice for buildings*, 3rd edn, London: ICE Publishing, p. 312.

3 Pietra, D., Pampanin, S., Mayes, R.L. *et al.*, 2014, 'Design of base-isolated buildings: an overview of international codes', Proceedings of the New Zealand Society for Earthquake Engineering Conference, Wellington, paper O47.

4 Martelli, A., Clemente, P., De Stefano, A., Forni, M. *et al.*, 2013, 'Development and application of seismic isolation energy dissipation and other vibration control techniques in Italy for the protection of civil structures, cultural heritage and industrial plants', Proceedings of the 13th World Conference on Seismic Isolation, Energy Dissipation and Active Vibration Control of Structures, Sendai, p. 1.

5 Becker, T.C., Furukawa, S., Mahin, S.A. & Nakashima, M., 2010, 'Comparison of US and Japanese codes and practices for seismically isolated buildings', Proceedings of the 2010 ASCE Structures Congress, pp. 2330–8, p. 2331.

6 Pan, P., Zamfirescu, D., Nakashima, M. *et al.*, 2005, 'Base-isolation design practice in Japan: introduction to the post-Kobe approach', *Journal of Earthquake Engineering*, vol. 9, no. 1, pp. 147–71, p. 162.

7 Feng, D., Miyama, T., Liu, W. & Chan, T., 2012, 'A new design procedure for seismically isolated buildings based on seismic isolation codes worldwide', Proceedings of the 15th World Conference on Earthquake Engineering, Lisbon.

8 Cutfield, M.R., Ma, Q.T. & Ryan, K.L., 2014, 'Cost-benefit analysis of base isolated and conventional buildings: a case study', Proceedings of the New Zealand Society for Earthquake Engineering Conference, Wellington, paper O46, p. 7.

9 *ASCE 7 2009: Minimum design loads for buildings and other structures*, American Society of Civil Engineers, Reston VA., Chapter 17 Seismic design requirements for seismically isolated structures.

10 Ozdemir, G. & Akyuz, U., 2010, 'Response of isolated RC buildings under bi-directional near-fault ground motions', Proceedings of the 9th US National and 10th Canadian Conference on Earthquake Engineering, Toronto, paper 460.

11 Kircher, C.A., 2006, 'Chapter 11 seismically isolated structures', in Building Seismic Safety Council, *NEHRP recommended provisions: design examples - FEMA 451*, National Institute of Building Sciences, Washington, D.C., p. 11–10.

12 *How to plan and implement seismic isolation for buildings*, 2013, The Japan Society of Seismic Isolation, Tokyo: Ohmsha, p. 46.

13 Saiki, K., Kitamura, Y., & Kani, N., 2013 'Damage of expansion-joints for seismically isolated buildings and countermeasures', Proceedings of the 13th World Conference on Seismic Isolation, Energy Dissipation and Active Vibration Control of Structures, Sendai.

14 *ASCE 7 2009: Minimum design loads for buildings and other structures*, American Society of Civil Engineers, Reston VA., Proposed 2014 changes to 'Chapter 17 Seismic design requirements for seismically isolated structures', C 17.2.6.

15 Ko, E., Morgan, T., Bello, M., Bailey, R. *et al.*, 2010, 'Base isolated structure – the new San Francisco general hospital & trauma center', Proceedings of the 9th US National and 10th Canadian Conference on Earthquake Engineering, Toronto, paper 1378.

Economics of seismic isolation

Introduction

In the usually cost-driven construction sector, it is particularly important for cost differences associated with seismic isolation to be defined. Unfortunately, as we explain below, it is impossible to give readers an appreciation of how the cost of a seismically isolated building differs from that of a conventional building with a high degree of accuracy due to the large number of factors at play. However, first we must check we are comparing like with like. Comparing the construction costs of these two types of buildings is like comparing the cost of an ordinary motor vehicle to one with advanced safety features. One vehicle has an improved braking system to avoid damage in the first place, and then, in the event of a crash, energy-absorbing crush zones in the chassis and air bags to increase protection. There is no doubt about the improved safety performance for which we should expect an increased price.

Few studies have compared the initial costs of seismically isolated buildings to those of conventional buildings *with the same seismic performance*. Unfortunately, special buildings, often housing essential facilities, rather than typical buildings are those that are reported upon. Furthermore, the studies don't capture the reduction in downtime and other benefits seismic isolation provides. Nevertheless, Ron Mayes reports how the construction cost of a nuclear power plant was 2 per cent less expensive, and a fire command and control facility 6 per cent cheaper with seismic isolation.[1] Seismic isolation led to lower costs to brace piping and electrical and mechanical equipment which lowered the overall cost. In a more recent example, Ko reports that a major new seismically isolated hospital was cheaper and easier to build than a conventional hospital building. Because the code places limits on the inter-storey drifts of hospitals, or in other words, how much further one storey moves horizontally compared to the storey below, the savings of steel framing more than offset the costs of isolation.[2] Other advantages he mentions are the reduced floor accelerations and therefore less damage to contents, and shallower beams and smaller columns that improve space planning.

Other reasons for a lack of definitive seismic isolation cost information is that construction economics vary from country to country, and that most seismically isolated buildings are different. Perhaps site seismicity or soil conditions vary, or the isolation plane is in a different position. In spite of all these differing factors some boundaries can be drawn around cost variations between conventional and isolated buildings. Drawing from eight case-studies mainly in the USA, Mayes reports that the cost of construction increases up to 5 per cent and savings of up to 3 per cent where using seismic isolation.[3] Japanese experience is summarized: 'Generally, for a building with less than about ten stories, the initial construction cost is several per cent higher than for the building without isolation, but for structures more than ten stories, there is almost no difference in construction cost.'[4] This 2013 information suggests there has been a reduction in seismic isolation costs since 1999. Then the cost premium for seismic isolation was estimated at 7.6 per cent on the basis of survey responses of 25 Japanese designers and 24 owners and researchers.[5] It is likely that the costs of seismic isolation will continue to fall as more bearing manufacturers enter the market and regulatory requirements are rationalized. In New Zealand, a study of four isolated hospitals that were completed between 2005 and 2013 revealed that the total additional cost of seismic isolation was 3 per cent of their construction costs.[6] Since hospitals are heavily serviced and therefore cost more per square metre, the additional isolation costs for most other building types can be expected to be slightly greater. In a European example involving the fast-track build of 4500 apartments after the damaging 2009 L'Aquila earthquake, Calvi reports that seismic isolation represented 2 per cent of the total cost.[7] We stress once again that these cost comparisons neglect completely the potentially enormous cost benefits that seismic isolation delivers after a major earthquake by reducing injuries and downtime.

Additional and reduced isolation construction costs

If there are additional costs of protecting a building and its contents by seismic isolation, how do they arise? Almost all of the additional costs are incurred in the vicinity of the isolation plane. This means the cost of isolation per square metre of construction is reduced both by reducing the ground floor plan area and increasing the numbers of storeys. Additional costs arise from a possible additional suspended floor, isolation devices such as bearings, provision of moats or rattle-space including retaining walls, and moat covers which are usually required unless the isolation plane is above or at ground level. Movement joints between adjacent buildings add to the costs, as well as detailing of flexible electrical and other services, stairs and elevators that cross the isolation plane, and increased design and peer review fees. To these costs could be added the loss of income from being unable to develop as much of the site area as normal. This is due to wider-than-normal movement gaps above the isolation plane adjacent to neighbouring buildings and site boundaries, and is rarely acknowledged.

Of all these costs, the most significant is the cost of an additional suspended floor immediately above the isolation plane. Ryan and others highlight this in their breakdown of the additional costs of seismic isolation for a three-storey steel framed US building (Table 9.1).[8]

Table 9.1 Detailed additional seismic isolation construction costs

Item	*% of additional cost*
Excavation	8
Retaining wall and moat cover	11
Isolator pedestals	1
Isolation devices	27
Level 1 floor and framing	36
Flexible connections	6
Crawl space drainage and lighting	6
Suspended elevator shafts	5

In the study of the four New Zealand seismically isolated hospitals mentioned previously, the costs of isolators ranged from 1–2 per cent of the construction cost including the cost of isolator prototype tests, and all other costs, including the rattle space and retaining walls lay within the same range.[9] In all cases, basements at least under part of the ground floors of the hospitals are used for mechanical services or car parking.

The additional costs of seismic isolation discussed above are inevitable. Of course, the percentage contribution each item adds to the final cost varies from project to project. However, as already mentioned, there are examples where seismic isolation is cost-neutral or even cheaper than conventional construction. These situations occur most commonly in essential facilities such as hospitals, civil defence offices and fire stations – operations that must function post-earthquake without interruption. The conventional approach to the design of these facilities is to increase their structural strength and stiffness. At least in New Zealand, such buildings are almost twice as strong as usual, their structural members significantly larger and more expensive. Since seismic isolation may meet and exceed the required seismic performance without the need of larger structure, some cost savings are expected.

Even for seismically isolated buildings with normal occupancy, such as apartment or office buildings, some structural savings will partially off-set the additional costs of seismic isolation. Structural components, such as beams, are not required to withstand seismic overload and absorb earthquake energy by yielding of either structural steel sections or reinforcing steel embedded in concrete. Columns may be able to be reduced in strength and structural detailing reduced and simplified. This means fewer reinforcing ties in reinforced concrete columns and beams, or simpler beam to column connections in structural steel frames, speeding up construction and lowering cost.

Cost savings other than to primary structural framing are likely as well. In isolated buildings, mechanical and electrical plant and architectural elements, such as suspended ceilings, require less bracing in order to prevent overturning, sliding or just resist their own earthquake inertia forces. Also, because horizontal superstructure deflections are reduced by seismic isolation, savings arise from simpler and smaller movement details. According to a structural engineer of a new office building in Christchurch, seismic isolation 'had a significant benefit when we were detailing the cladding system and attachments for the curved façades. If we had used a conventional structure design, then the inter-storey drifts would have made the curved glass façades cost-prohibitive'(Figure 9.1).[10]

9.1 Office building in Christchurch where reduced horizontal inter-storey movement as a result of seismic isolation reduced the costs of the curved wall glazing

Source: Mark Southcombe

Pre-earthquake economics

EVEN before the benefits from seismic isolation are realized after an earthquake, seismic isolation can positively affect a building owner's income. For example, Japanese apartment dwellers are prepared to pay higher rentals for the security and peace of mind resulting from superior seismic performance. The other possible on-going financial advantage of seismic isolation is earthquake insurance. Not only is it easier to obtain insurance cover in a market insurers may be withdrawing from, such as post-earthquake Christchurch and other higher risk regions, but significant cost savings are also possible. Armed with an engineering report of his or her isolated building, an owner might succeed in negotiating a significant reduction to the annual earthquake insurance premium. Alternatively, an owner can self-insure. It is most unlikely the damage to an isolated building would exceed a typical insurance deductible of approximately 5 per cent of the sum insured (building and contents). The deductible and savings on annual premiums will more than pay for any additional cost of seismic isolation. Earthquake insurance premiums fluctuate considerably. They depend upon many factors including the seismicity of a site and the state of the insurance industry at that particular time, but when they are of the order of 1 per cent of the sum insured, self-insurance becomes very attractive. One significant impediment to self-insurance is the typical requirement of financial lenders, such as banks, to require full insurance cover, but there is at least one US precedent of a lender agreeing to waive earthquake insurance on the basis of the enhanced seismic performance provided by seismic isolation.[11]

Post-earthquake economics

THE most significant economic benefits of seismic isolation occur during and immediately after a major earthquake. Occupants' personal losses are reduced, together with costs arising from potential deaths and injuries. Provided that the building is not damaged by adjacent buildings, is accessible and inhabitable due to continued supply of water, sewerage and energy, then losses of rent, disruption of production and loss of market share are minimized to the greatest possible extent. Although insurance companies offer business disruption insurance, seismic isolation provides by far the most elegant solution to the problem, through avoidance.

Owners of conventional buildings must assess these possible losses for themselves. They are often huge and difficult to predict accurately. For example, how much time might be spent accessing, planning and repairing a building in the aftermath of a major earthquake? Given the extent of post-earthquake community disruption, very little works as normal. Everything is in a state of flux, including local authorities. In Cutfield and Ma's study, business interruption or downtime cost was estimated at five to ten times the cost of building damage, unless the owner was able to re-locate quickly and efficiently.[12] Based on his experience following the 2010–11 Canterbury earthquake sequence, which caused so much destruction in Christchurch, David Whittaker calculates the cost of business disruption as the rental income lost for 3 years – equivalent to losing 30 per cent of the building value.[13] For high-tech manufacturers or research establishments, far higher levels of business disruption losses can be imagined, possibly leading to businesses failing.

Life-cycle analyses

We all know the pitfalls associated with choices made solely on the basis of first or initial costs. The cheapest cladding system may look fine during installation, but what's it like five or ten years later? Life-cycle analyses are particularly useful and revealing for simple and relatively quantifiable situations like this. But when a client asks about the life-cycle costs or savings from seismic isolation, the calculations are not straight forward. Even if sophisticated Net Present Value methods are adopted, the answers depend upon fundamental assumptions that are subject to large variability. However, all published life-cycle analyses for buildings located in high seismic hazard areas of western North America and New Zealand consistently demonstrate the cash flow benefits of seismic isolation.[14, 15, 16, 17] For example, Terzic and others calculate a minimum 3 per cent return on investment over a 50-year time frame.[18] They exclude potential insurance savings and acknowledge even more attractive returns for seismic isolation if downtime estimates are refined.

Most life-cycle analyses neglect the costs and benefits of earthquake insurance. However, in his simplified analysis based upon annualised costs, Whittaker makes a compelling case for owners of seismically isolated buildings to self-insure.[19] He shows that the annual costs of earthquake damage plus business interruption, insurance premium assuming no reduction for the isolated building, and annualised deductible are similar for owners of conventional and seismically isolated buildings. The spike in insurance premiums immediately after the Canterbury earthquake has reduced dramatically from approximately 1 per cent to 0.25 per cent. But an update of Whittaker's spreadsheet shows that the annualised cost of an uninsured (self-insured) isolated building is only 7 per cent of an insured conventional building.

Notes

1 Mayes, R.L., 1990, 'The economics of seismic isolation in buildings', *Earthquake Spectra*, vol. 6, no. 2, pp. 245–63.

2 Ko, E., Morgan, T., Bello, M., Bailey, R. *et al.*, 2010, 'Base isolated structure – the new San Francisco general hospital & trauma center', Proceedings of the 9th US National and 10th Canadian Conference on Earthquake Engineering, Toronto, paper 1379.

3 Mayes, R.L., 1990, 'The economics of seismic isolation in buildings', *Earthquake Spectra*, vol. 6, no. 2, p. 252.

4 The Japan Society of Seismic Isolation (ed.) 2013, *How to plan and implement seismic isolation for buildings*, Tokyo: Ohmsha, p. 38.

5 Clark, P.W., Aiken, I.D., Nakashima, M., Miyazaki, M. & Midorikawa, M., 2000, *New design technologies: the 1995 Kobe (Hyogo-ken Nanbu) earthquake as a trigger for implementing new seismic design technologies in Japan*, Lessons learned over time, Learning from earthquakes, vol. 3, Earthquake Engineering Research Institute, Oakland, California.

6 Charleson, A.W. & Allaf, N.J., 2012, 'Costs of base-isolation and earthquake insurance in New Zealand', Proceedings of the New Zealand Society for Earthquake Engineering Conference, Christchurch, Paper no. 041, p. 6.

7 Calvi, G.M., 2010, 'L'Aquila earthquake 2009: reconstruction between temporary and definitive', Proceedings of the New Zealand Society for Earthquake Engineering Conference, Wellington, paper 01, p. 11.

8 Ryan, K.L., Sayani, P.J., Dao, N.D., Abraik, E. & Baez, Y.M., 2010, 'Comparative life cycle analysis of conventional and base-isolated building theme buildings', Proceedings of the 9th US and 10th Canadian Conference on Earthquake Engineering, Toronto, p. 3.

9 Charleson, A.W. & Allaf, N.J., 2012, 'Costs of base-isolation and earthquake insurance in New Zealand', Proceedings of the New Zealand Society for Earthquake Engineering Conference, Christchurch, paper no. 041, p. 6.

10 Kane, R., 2014, 'Base isolation sets new standard in Christchurch', *Building Today*, vol. 24, no. 10, p. 28.

11 Comartin, C., 2003, *Earthquake insurance*, viewed 11 April 2011, www.comartin-reis.com/News/Eq%20 Ins.htm

12 Cutfield, M.R., Ma, Q.T. & Ryan, K.L., 2014, 'Cost-benefit analysis of base isolated and conventional buildings: a case study', Proceedings of the New Zealand Society for Earthquake Engineering Conference 2014, Auckland, paper no. O46, p. 7.

13 Whittaker, D., 2012, 'Economic benefits of seismic isolation', Proceedings of the SESOC New Zealand Conference, Auckland, p. 3.

14 Ryan, K.L., Sayani, P.J., Dao, N.D., Abraik, E. & Baez, Y.M., 2010, 'Comparative life cycle analysis of conventional and base-isolated buildings', Proceedings of the 9th US and 10th Canadian Conference on Earthquake Engineering, Toronto, p. 3.

15 Whittaker, D., 2012, 'Economic benefits of seismic isolation', Proceedings of the SESOC New Zealand Conference, Auckland.

16 Cutfield, M.R., Ma, Q.T. & Ryan, K.L., 2014, 'Cost-benefit analysis of base isolated and conventional buildings: a case study, Proceedings of the New Zealand Society for Earthquake Engineering Conference, Auckland, paper no. O46, p. 7.

17 Terzic, V., Merrifield, S.K. & Mahin, S.A., 2012, 'Lifecycle cost comparisons for different structural systems designed for the same location', Proceedings of the 15th World Conference on Earthquake Engineering, Lisbon.

18 Ibid. p. 9.

19 Whittaker, D., 2012, 'Economic benefits of seismic isolation', Proceedings of the SESOC New Zealand Conference, Auckland, p. 3.

– 10 –

Construction and maintenance

Introduction

ALTHOUGH the construction of seismically isolated and conventional buildings is similar, two aspects related to isolation systems require special attention during construction. First, since the detailing of an isolation system in the vicinity of its isolation plane is unusual, designers need to put extra effort into communicating and explaining these different features and requirements to the contractor. Second, unlike conventional construction, isolated buildings require the insertion of various isolation devices.

Once constructed, a seismically isolated building has unique maintenance requirements. Just like elevator and fire alarm systems need regular inspection and maintenance, so also isolation systems. The consequences of an isolation system being compromised, say by building materials blocking a seismic movement gap, or bearings suffering severe corrosion, are very serious. The safety of the entire building is at risk.

Construction

SINCE most contractors haven't built a seismically isolated building before, its designers need to explain how the isolation system works, and its construction implications. Although some of this communication can be achieved through contract documentation, there is no substitute for designers requesting special meetings to outline the unique aspects of the isolation system. During the construction of an isolated building, the contractor has to act against his or her instincts. A contractor is used to fixing or joining elements together rather than letting them slide, and is used to filling gaps rather than leaving them open. Unless explained clearly and perhaps repeatedly, habitual ways of building, like anything else in life, are difficult to change. For this reason thorough construction inspections are recommended, possibly by a seismic isolation construction supervisor.[1]

Examples of construction mistakes affecting the performance of seismically isolated buildings have been reported. Kani and others state that 25 per cent of the seismically isolated buildings they inspected after the 2011 Tōhoku earthquake suffered damage to moat covers and other components of the isolation systems.[2] Some of this damage was due to faulty or careless construction. During post-earthquake inspections of the Christchurch Women's Hospital in 2010 and 2011, buckling damage to a length of seismic moat cover at ground level, and damage to an air bridge connecting the hospital to another building was evidence of how the effectiveness of the isolation system was reduced.[3, 4]

Another aspect of seismic isolation that is new to most contractors is the procurement process of the isolation devices. Early ordering of isolation devices to avoid installation delays and the need for temporary inserts is highly recommended. Between three to six months should be allowed for device manufacture, excluding transportation time. The client may wish to order and purchase devices directly and have them delivered to site in order to avoid the contractor's margin. The structural engineer plays important roles during procurement, depending on which procurement strategy is adopted. Does the engineer specify a complying system, specify performance requirements with device vendors undertaking the design, or is a complying system specified along with its performance requirements that can then be matched by other devices?[5] The final decision has to balance the need for optimal design, to include all potential bidders, and finally, to reduce the difficulty and effort of checking large numbers of analyses as part of the bid evaluation process.

During the final weeks of construction, work should commence on communicating to all relevant people the fact that a building is seismically isolated. Permanent notices warning against placing objects that might prevent unobstructed movement between isolated and non-isolated structure need to be posted. One notice should be placed at the entrance to the crawl space (if provided), and other notices placed around the building perimeter in an attempt to prevent any obstructions reducing the effectiveness of the seismic movement gap. Also at this time, a Building Seismic Isolation Manual should be prepared jointly by the structural engineer and architect. Written for the building owner, its purpose is to describe the isolation system, explain how it is expected to perform, and document all the seismic isolation details. Since the manual needs to be referred to during regular maintenance inspections, it should include design details and as-built drawings as necessary. All structural and non-structural isolation details, such as movement gap details, should be included to help future inspectors with no first-hand knowledge of the building.

Maintenance

PHYSICAL systems deteriorate with age. The oxygen and water that sustain us oxidize and corrode construction materials, including those of seismic isolation systems. The thin outer layer of a rubber bearing will suffer a loss of elasticity, while the surfaces of sliding joints are prone to sticking, or even locking up in the presence of moisture or surface debris. However, seismic isolation systems face a possibly more severe danger during their lifetimes – inadvertent human interference with seismic movement gaps. A gap filled in over just part of its length could cause severe structural damage to an isolated building during an earthquake. If the isolated structure can't move freely within its enclosing retaining walls then it is no longer isolated. It will be subject to inertia forces far larger than those designed for.

A maintenance programme therefore needs to address both the natural and human threats to which seismic isolation is vulnerable. The 2012 Japanese Society of Seismic Isolation maintenance standard specifies the frequency and nature of inspections of seismic isolation devices, such as bearings, sliders and dampers; the seismic isolation plane including the perimeter details of seismic clearance gaps and movement joints; and the flexible services pipes and wiring entering the building.[6] The standard also defines five types of inspection:

- inspections at construction completion;
- periodic inspections consisting of visual annual inspections except for five years after construction and every ten years thereafter when certain measurements are taken (including devices removed and tested);
- emergency inspections after felt earthquakes, severe wind storms, floods or fires;
- detailed inspections to follow up defects discovered during periodic inspections; and
- inspections after renovation or repairs to the isolation system such as to flexible pipes and movement joints. Inspections after any interior construction in the vicinity of the isolation plane, or any construction outside the building close to its perimeter, including landscaping work should also be included in this category.

After the 2011 Tōhoku earthquake the lead dampers in at least one building cracked and needed replacement.[7]

Personal experience of one of the authors of this book has also shown the need for this type of maintenance regime. Before the mistake was rectified, an isolated building being refurbished and subject to landscaping changes was observed to be strongly tethered to the surrounding ground by horizontal steel bars. The bars supported a fence that was part of a landscaping project abutting the isolated building.

Kelly and others cite a similar but more serious example:

> At the LA County Fire Command Center, the contractor had poured a reinforced concrete slab under the floor tiles at the main entrance to the buildings, preventing free movement in the E-W direction. Apparently the reinforcing was added after the contractor had replaced the tiles several times after minor earthquakes and did not realize that this separation was designed to occur.[8]

Examples of the importance of periodic inspections include the need to replace the rubber bearings of what some regard as the first base-isolated building in the world. The 40-year-old rubber block bearings (without the benefit of the steel confining plates of modern bearings) had cracked and deformed.[9] Twenty-four years after their installation, the lead-extrusion dampers at the Wellington Central Police station required maintenance. Tests showed that an undesirable bond between the lead and the steel cylinders had formed. The interface between the materials needed re-lubricating to maintain the safety of the building.[10]

The maintenance programme for a seismically isolated building should ideally be established and agreed to by the client before construction is complete. We believe that the programme should be as legally binding as those for elevators and fire alarm systems that are regularly inspected and maintained. In New Zealand, most non-residential buildings require an annual Building Warrant of Fitness. It consists of a statement supplied by the owner to the local council stating that the building systems listed in a compliance schedule have been checked and maintained in accordance with the schedule for the previous 12 months. The compliance schedule, which should include the seismic isolation system, is agreed to by the council at the time the building owner applies for Building Consent to begin construction.[11] If other countries do not have such a system, it is the responsibility of the structural engineer to set up a fail-safe inspection and maintenance system. It needs to outlive the structural engineer and the existence of his or her firm for the entire duration of the life of the building. Anything less cannot guarantee the effectiveness and safety of a seismic isolation system.

Notes

1 The Japan Society of Seismic Isolation (ed.) 2013, *How to plan and implement seismic isolation for buildings*, Tokyo: Ohmsha, p. 62.

2 Kani, N., Ogino, N., Kitamura, Y. & Kitamura, H., 2012, 'Effects of seismically isolated buildings during the huge 2011 earthquake in Japan', Proceedings of the 15th World Conference on Earthquake Engineering, Lisbon. p. 1.

3 Gavin, H.P. & Wilkinson, G., 2010, 'Preliminary observations of the effects of the 2010 Darfield earthquake on the base-isolated Christchurch Women's Hospital', *Bulletin of the New Zealand Society for Earthquake Engineering*, vol. 43, no. 4, pp. 360–7.

4 Kuang, A., Sridhar, A., Gavin, H. & Gutschmidt, S., 2013, 'Analysis of the seismic response of the Christchurch Women's Hospital', Proceedings of the 2013 New Zealand Society for Earthquake Engineering Conference, p. 1.

5 Kelly, T.E., Skinner, R.I. & Robinson, W.H., 2010, *Seismic isolation for designers and structural engineers*, Kanpur: NICEE, p. 115.

6 Kani, N., 2013, 'Maintenance standards for seismic-isolation buildings', Proceedings of the 13th World Conference on Seismic Isolation, Energy Dissipation and Active Vibration Control of Structures, Sendai, Japan, p. 2.

7 Motosaka, M. & Mitsuji, K., 2012, 'Building damage during the 2011 Pacific Coast Tohoku earthquake', *The Japanese Geotechnical Society Soils and Foundations*, vol. 52, no. 5, pp. 929–44, p. 931.

8 Kelly, T.E., Skinner, R.I. & Robinson, W.H., 2010, *Seismic isolation for designers and structural engineers*, Kanpur: NICEE, p. 327.

9 Gjorgjieiv, I. & Garevski, M., 2012, 'Replacement of the old rubber bearings of the first base isolated building in the world', Proceedings of the 15th World Conference on Earthquake Engineering, Lisbon.

10 Smart, C.R., 2013, 'Wellington Central Police Station base isolation maintenance', *Bulletin of the New Zealand Society for Earthquake Engineering*, vol. 46, no. 3, pp. 141–56.

11 Wellington City Council 2005, *Building warrant of fitness*, viewed 15 December 2014, http://wellington.govt.nz/services/consents-and-licences/building-consents/building-warrant-of-fitness

– 11 –

Conclusions

Confidence in seismic isolation

OVER the last five years hundreds of seismically isolated buildings have performed very well in damaging earthquakes. This has confirmed the results of thousands of computer analyses and numerous full-scale physical earthquake simulations of isolated buildings and isolation devices. As a consequence of positive comparable results from field performance, laboratory testing and computer modelling, we now can have unprecedented confidence in the improved performance seismic isolation can provide.

Benefits of seismic isolation

THE potential benefits of seismic isolation are enormous. The inclusion of the word 'potential' acknowledges the likelihood, in the range of 80–90 per cent, that the design earthquake will *not* occur during the life of a building. However, during and after the design earthquake striking an isolated building, the building owner will benefit in many different ways from the protection provided by seismic isolation. It is not too dramatic to say that lives, jobs and businesses could be saved. And even before the 'big one' strikes, if ever, financial, aesthetic and functional benefits arise from seismic isolation.

Design freedom and limitations

THE case studies in Chapter 6 illustrate how seismic isolation can offer architects opportunities to achieve and reinforce certain design concepts and architectural qualities. As designers, and especially structural engineers, become even more confident in the technology, seismic isolation may encourage structural forms in seismic zones that are more irregular than those recommended for conventional construction. In spite of this freedom, the discipline of careful architectural detailing, particularly around movement gaps, needs constant attention.

Just as personal human freedom has its limits, so too does seismic isolation. Some seismological and geological conditions, and urban building sites are unsuitable for seismic isolation.

Uncertainties

SEISMIC isolation is a relatively new technology. Its 35 years of age represents one third of the era of modern earthquake engineering. As such, it is not surprising some technical aspects, such as the width of moat gaps, are still being debated. Although the science of seismic isolation is now well understood, some uncertainties linger related to the nature of and the characteristics of earthquake shaking. We are designing for a most unpredictable hazard. Also, post-construction uncertainties, such as how environmental conditions affect isolation devices and how building use, like the inadvertent blocking of movement gaps, might negate seismic isolation systems need to be reduced, or better, eliminated, by rigorous inspection and maintenance regimes.

Savings and additional costs of seismic isolation

MOST cost comparisons between conventional construction and seismically isolated buildings are misleading since the far improved seismic performance from seismic isolation is excluded. Possible significant savings in insurance premiums, especially if building owners self-insure, are not factored in. Even life-cycle cost analyses, that include all quantifiable losses including business disruption costs, exclude many intangible benefits of seismic isolation.

Seismic isolation can deliver construction savings for buildings housing essential facilities, such as hospitals, and modest increases in cost, usually less than 5 per cent can be expected for other building types. Published life cycle analyses show the economic value of seismic isolation.

Looking ahead

THIS book presents a state-of-the-art summary of the theory and practice of seismic isolation from an architectural perspective. Given its sole emphasis on recent and current practice, it is now appropriate to consider the future of seismic isolation. How might it develop in the coming years? Will it become more widely adopted and what might be the shapes of its changing applications?

The extent to which it will be adopted in the future depends on two major factors. The first is the frequency and severity of future damaging earthquakes. As witnessed after every large earthquake in an urban setting, the reality of human casualties and collapsed and badly damaged buildings leads to a surge in seismically isolated buildings. There is no reason why this response to earthquake damage will cease. Societies are becoming more risk adverse. We are aware of how the motor vehicle industry continues to enhance the safety features of cars. Based upon past experience, we predict that after every large earthquake, building owners will reflect on the vulnerability of their own buildings and the possibility of minimizing their future losses through seismic isolation.

The second factor that will influence the extent of the adoption of seismic isolation is the quality of the performance of seismically isolated buildings. If they continue to out-perform conventional construction, building owners will be inclined to adopt seismic isolation. If, however, one or more failures of isolation systems result in severe building damage, then seismic isolation will be passed-by and less effective seismic resisting approaches adopted. It is the responsibility of all parties with an interest in seismic isolation to not only maintain current standards of design, construction and maintenance, but to continually improve them. Over-confidence in this still relatively new technology must be avoided.

Apart from these two primary definers of the future of seismic isolation there are other drivers as well, all of which point towards building owners finding seismic isolation increasingly attractive. Structural design procedures are undergoing rationalisation and simplification as isolation technology matures. The insurance industry is becoming better acquainted with the benefits of seismic isolation, and on-going research and development is likely to lead to lower-cost isolation devices, passive as well as active. All these advances, as well as more building owners, architects and engineers requesting or recommending the application of seismic isolation in their buildings, will result in safer buildings and communities.

- INDEX -

For Product Safety Concerns and Information please contact our EU representative GPSR@taylorandfrancis.com Taylor & Francis Verlag GmbH, Kaufingerstraße 24, 80331 München, Germany

Batch number: 10777156

Printed by Printforce, the Netherlands